Palgrave Studies in the History of Social Movements

Series Editors
Stefan Berger
Institute for Social Movements
Ruhr University Bochum
Bochum, Germany

Holger Nehring
Contemporary European History
University of Stirling
Stirling, UK

Around the world, social movements have become legitimate, yet contested, actors in local, national and global politics and civil society, yet we still know relatively little about their longer histories and the trajectories of their development. This series seeks to promote innovative historical research on the history of social movements in the modern period since around 1750. We bring together conceptually-informed studies that analyse labour movements, new social movements and other forms of protest from early modernity to the present.

We conceive of 'social movements' in the broadest possible sense, encompassing social formations that lie between formal organisations and mere protest events. We also offer a home for studies that systematically explore the political, social, economic and cultural conditions in which social movements can emerge. We are especially interested in transnational and global perspectives on the history of social movements, and in studies that engage critically and creatively with political, social and sociological theories in order to make historically grounded arguments about social movements. This new series seeks to offer innovative historical work on social movements, while also helping to historicise the concept of 'social movement'. It hopes to revitalise the conversation between historians and historical sociologists in analysing what Charles Tilly has called the 'dynamics of contention'.

More information about this series at
http://www.palgrave.com/gp/series/14580

Claas Kirchhelle

Bearing Witness

Ruth Harrison and British Farm Animal Welfare (1920–2000)

Claas Kirchhelle
University College Dublin
Dublin, Ireland

ISSN 2634-6559 ISSN 2634-6567 (electronic)
Palgrave Studies in the History of Social Movements
ISBN 978-3-030-62791-1 ISBN 978-3-030-62792-8 (eBook)
https://doi.org/10.1007/978-3-030-62792-8

Cover illustration: Sidney Wheeler / Stockimo / Alamy Stock Photo

This Palgrave Macmillan imprint is published by the registered company Springer Nature Switzerland AG.
The registered company address is: Gewerbestrasse 11, 6330 Cham, Switzerland

Für Clara

Series Editors' Preface

Around the world, social movements have become legitimate, yet contested, actors in local, national, and global politics and civil society, yet we still know relatively little about their longer histories and the trajectories of their development. Our series reacts to what can be described as a recent boom in the history of social movements. We can observe a development from the crisis of labour history in the 1980s to the boom in research on social movements in the 2000s. The rise of historical interests in the development of civil society and the role of strong civil societies as well as non-governmental organisations in stabilising democratically constituted polities has strengthened the interest in social movements as a constituent element of civil societies.

In different parts of the world, social movements continue to have a strong influence on contemporary politics. In Latin America, trade unions, labour parties, and various left-of-centre civil society organisations have succeeded in supporting left-of-centre governments. In Europe, peace movements, ecological movements, and alliances intent on campaigning against poverty and racial discrimination and discrimination on the basis of gender and sexual orientation have been able to set important political agendas for decades. In other parts of the world, including Africa, India, and South East Asia, social movements have played a significant role in various forms of community building and community politics. The contemporary political relevance of social movements has undoubtedly contributed to a growing historical interest in the topic.

Contemporary historians are not only beginning to historicise these relatively recent political developments; they are also trying to relate them to a longer history of social movements, including traditional labour organisations, such as working-class parties and trade unions. In the longue durée, we recognise that social movements are by no means a recent phenomenon and are not even an exclusively modern phenomenon, although we realise that the onset of modernity emanating from Europe and North America across the wider world from the eighteenth century onwards marks an important departure point for the development of civil societies and social movements.

In the nineteenth and twentieth centuries the dominance of national history over all other forms of history writing led to a thorough nationalisation of the historical sciences. Hence social movements have been examined traditionally within the framework of the nation state. Only during the last two decades have historians begun to question the validity of such methodological nationalism and to explore the development of social movements in comparative, connective, and transnational perspective taking into account processes of transfer, reception, and adaptation. Whilst our book series does not preclude work that is still being carried out within national frameworks (for, clearly, there is a place for such studies, given the historical importance of the nation state in history), it hopes to encourage comparative and transnational histories on social movements.

At the same time as historians have begun to research the history of those movements, a range of social theorists, from Jürgen Habermas to Pierre Bourdieu and from Slavoj Žižek to Alain Badiou as well as Ernesto Laclau and Chantal Mouffe to Miguel Abensour, to name but a few, have attempted to provide philosophical-cum-theoretical frameworks in which to place and contextualise the development of social movements. History has arguably been the most empirical of all the social and human sciences, but it will be necessary for historians to explore further to what extent these social theories can be helpful in guiding and framing the empirical work of the historian in making sense of the historical development of social movements. Hence the current series is also hoping to make a contribution to the ongoing dialogue between social theory and the history of social movements.

This series seeks to promote innovative historical research on the history of social movements in the modern period since around 1750. We bring together conceptually informed studies that analyse labour movements, new social movements, and other forms of protest from early modernity to the present. With this series, we seek to revive, within the context of historiographical developments since the 1970s, a conversation between historians on the one hand and sociologists, anthropologists, and political scientists on the other.

Unlike most of the concepts and theories developed by social scientists, we do not see social movements as directly linked, a priori, to processes of social and cultural change and therefore do not adhere to a view that distinguishes between old (labour) and new (middle-class) social movements. Instead, we want to establish the concept 'social movement' as a heuristic device that allows historians of the nineteenth and twentieth centuries to investigate social and political protests in novel settings. Our aim is to historicise notions of social and political activism in order to highlight different notions of political and social protest on both left and right.

Hence, we conceive of 'social movements' in the broadest possible sense, encompassing social formations that lie between formal organisations and mere protest events. But we also include processes of social and cultural change more generally in our understanding of social movements: this goes back to nineteenth-century understandings of 'social movement' as processes of social and cultural change more generally. We also offer a home for studies that systematically explore the political, social, economic, and cultural conditions in which social movements can emerge. We are especially interested in transnational and global perspectives on the history of social movements and in studies that engage critically and creatively with political, social, and sociological theories in order to make historically grounded arguments about social movements. In short, this series seeks to offer innovative historical work on social movements, while also helping to historicise the concept of 'social movement.' It also hopes to revitalise the conversation between historians and historical sociologists in analysing what Charles Tilly has called the 'dynamics of contention.'

Claas Kirchhelle's *Bearing Witness: Ruth Harrison and British Farm Animal Welfare (1920–2000)* is an extremely readable and fascinating account of one of the most well-known animal rights' activists in Britain in the twentieth century who has been, at the same time, strangely neglected to date by academic research. She has been an inspiration to generations of other activists and the opposition to 'factory farming'—her concept—would be unthinkable without her path-breaking work. Born into a family of Edwardian radicals practicing vegetarianism, pacifism, feminism, and socialism, Harrison was a life-long reformist campaigner for animal rights believing in the power of social movements to achieve change for the better in democratically constituted societies.

Kirchhelle's book is outstanding not only in tracing Harrison's biography, uncovering a wealth of new material from archives and through interviewing a range of fellow activists and others who had an intimate knowledge of Harrison's life and work. The author is also adept at relating the animal rights activism of Harrison and her associates with wider ethical and social concerns that were prominently discussed in twentieth-century Britain, such as wider environmental concerns and questions surrounding animal-human relationships.

Harrison is perhaps best known for her work *Animal Machines* that she published in 1964. It catapulted her to international fame and put her centre-stage in the civic activism surrounding animal rights' issues. Among all the social movements that have been studied in the twentieth century it would be fair to say that so far the animal rights' movement has not been the most prominent. Hence the book is also a call on social movement researchers to look more closely at a movement that has exceptional contemporary relevance but also deep historical roots.

Kirchhelle demonstrates how Harrison's activism was built around the twin pillars of moral improvement and welfare. The way in which humans treated animals was immoral and against the welfare of animals, but it also denigrated the humanity of humans who allowed such treatment of animals to be legal. Using animals for scientific research, for example, could only be justified when animals were treated humanely, because science itself had to be oriented towards humanism. Intensive food production could not be justified, if it was unethical. The Brambell Committee on Animal Welfare and its 1965 report built on the insights in Harrison's 1964 bestseller and was to have a major impact on future discussions about animal welfare in Britain.

What emerges as remarkable about Harrison is her ability to move between different generations and also to be active at one and the same time in establishment and anti-establishment circles. Her own self-styling as a 'lone wolf' allowed this precarious existence between camps usually seen as being at loggerheads with each other.

From the 1970s onwards the setting of welfare standards for animals became the most contested site for welfare activism and Kirchhelle traces the role of Harrison as a key figure mediating between scientists, legislators, and protesters. Working on various British and European welfare committees, Harrison campaigned actively to see several of her demands that she had made in her 1964 book, fulfilled, for example, the abolition of veal crates. She often took a middling position between the more radical demands of those seeing in the treatment of animals by humans nothing but 'speciest exploitation' and those who argued that there was a decisive difference between humans and animals but that it was incumbent on humans for religious and moral reasons to treat animals well. Overall Kirchhelle has provided us with an extremely insightful history of a key British animal rights' campaigner of the twentieth century whose actions and publications had a global ring and whose life hopefully will inspire others to take up the theme of social movements working on behalf of animal rights.

Bochum, Germany | Stefan Berger
Stirling, Scotland | Holger Nehring

Timeline

Ca. 1892	Birth Clara Birnberg
1893	Birth Samuel ('Simy') Weinstein
1910–1912	Birnberg attends Slade School of Art
1911	Samuel Weinstein joins Young Socialists Britain introduces Protection of Animals Act
1915	Isaac Rosenberg enlists and paints Clara Birnberg (*Girl in a Red Dress*)
1916	Samuel Weinstein, John Rodker, and Jonas Birnberg object to conscription. Weinstein and Rodker are imprisoned as conscientious objectors
Post-1918	The Weinsteins anglicise their names to Clare and Stephen Winsten
1920	Birth Ruth Harrison (*née Winsten*) in London
1926	Foundation UFAW
1939	Ruth Harrison enrols in Bedford College, London University Evacuation to Cambridge, where Ruth Harrison joins the Society of Friends and likely meets W.H. Thorpe
1940	The Winstens move to Ayot St Lawrence and become neighbours of George Bernard Shaw
1943–1945	Ruth Harrison works for the Friends' Ambulance Unit (FAU) as a nurse in Whitechapel, Lichfield, and Islington
1945–1946	Harrison aids Friends' Ambulance Unit relief efforts in the Ruhr area and Schleswig-Holstein (Germany)
1946–1948	Harrison enrols in the Royal Academy of Dramatic Art (RADA)
1949	Niko Tinbergen moves from Leiden to Oxford
1951	Festival of Britain

1954	Marriage to Dexter ('Dex') Harrison
	Rationing lifted in Britain
1957	Campaign for Nuclear Disarmament (CND) formed
1959	3Rs published by Russell and Burch
Ca. 1960	Crusade Against All Cruelty to Animals pushes leaflet under Ruth Harrison's door
1960	John Dugdale introduces Animals (control of intensified methods of food production) Bill
1961	Ruth Harrison begins work on *Animal Machines*
	Protests against 'field sports' at RSPCA meeting
1962	US publication of *Silent Spring* (UK publication 1963)
	Ruth Harrison contacts Rachel Carson in November
1963	Rachel Carson agrees to write foreword for *Animal Machines* in May
	The Observer agrees to publish articles on 'factory farming'
	Hunt Saboteurs Association formed
1964	Publication of *Observer* articles and *Animal Machines* in March
	MAFF decides to establish committee to review "intensive husbandry methods" in late March
	Death of Rachel Carson in April
1964–1965	Brambell Committee reviews intensive farming
1965	Brambell Report is published in December with annexe on four freedoms by W.H. Thorpe
1966	MAFF announces decision to enact a new welfare bill and establish FAWAC
1967	Establishment of FAWAC; Ruth Harrison is appointed
	Ruth Harrison registers Ruth Harrison Welfare Trust (soon renamed into Ruth Harrison Research Trust)
	MAFF announces outline of Agriculture (Miscellaneous Provisions) Bill
1968	First meeting of Ruth Harrison Research Trust
	FAWAC submits welfare code proposals to MAFF
1969	Ruth Harrison is appointed to RSPCA council in April
	Welfare codes are resubmitted to FAWAC
	Ruth Harrison leaks BFSS letter to the League Against Cruel Sports
1970	RSPCA Reform Group founded and members elected to RSPCA Council
	Richard Ryder coins 'speciesism'
1971	Publication of UK's first voluntary welfare codes

	Publication of *Animals Men and Morals* by members of the Oxford Group; contains a contribution by Ruth Harrison
	First Meeting RSPCA Farm Animal Livestock Committee
1972	RSPCA advisory expert committees on animal experimentation is founded
	West German Protection of Animals Act
	Election of Reform Group members to RSPCA Council
1973	Ruth Harrison loses libel suit against Nadia Nerina in May and is ordered to pay £30,000
	Nobel Prize in Physiology or Medicine for Tinbergen, Lorenz, and Frisch (ethology)
	First RSPCA conference on animal experimentation
	Arson attack on Hoechst Pharmaceuticals in Milton Keynes by the Band of Mercy (predecessor ALF)
1974	Partial enactment of FAWAC ban recommendations on the docking of cattle, the winging of and surgical castration of poultry
	Ruth Harrison Research Trust renamed into FACT
1974–1975	Reforms of RSPCA structure
1975	Personal bankruptcy of Ruth Harrison in April
	Publication of Peter Singer's *Animal Liberation*
1976	RSPCA votes to oppose hunting with hounds and refocuses campaigning to include farm animals
	Britain ratifies the European Convention for the Protection of Animals Kept for Farming Purposes
1977	FACT begins to support strawyard experiments by John Webster
	RSPCA Animal Rights Conference at Trinity College, Cambridge
1977–1979	Richard Ryder RSPCA Council chairmanship
1979	Foundation of FAWC; Ruth Harrison is appointed
1980	Foundation Eurogroup for Animals
1980s	Ruth Harrison attends European meetings of T-AP
1984	FAWC publishes 117 recommendations for animal welfare
1986	British ban of individually penned calf crates
	Ruth Harrison awarded an OBE
1987	Death Dexter Harrison
1988	Significant increase of FACT funding
1989	Death of Clare Winsten
1990	Veal crate ban is enacted
1991	Harrison retires from FAWC
	Death Stephen Winsten
1994	Launch RSPCA Freedom Foods

Ca. 1996	Cancer diagnosis Ruth Harrison
1999	Sow and tether stalls are banned in Britain
	Announcement of battery cage ban by 2012
	Ruth Harrison resigns FACT chairmanship in September
2000	Death Ruth Harrison in June
	November FACT Memorial Meeting at University of Westminster (200 letters of support and government, NGO, and press attendees)
2013	"Rachel Carson & Ruth Harrison: 50 Years on Conference" at Oxford University—launch of second edition of *Animal Machines*

Acknowledgements

Six years ago, I became interested in the life and work of Ruth Harrison while conducting my doctoral research on antibiotics. Although many authors have acknowledged the importance of *Animal Machines*, I was intrigued by the fact that none had focused on the wider life and work of its author and was surprised by how much there was to discover. What started as a side-enquiry into Harrison's life quickly turned into a full-blown book project on Harrison and the fascinating world she inhabited.

Many people have helped 'grow' this project. Prof. Mark Harrison patiently allowed me to become distracted by Ruth Harrison for prolonged periods of my PhD and read several drafts of the book. Dr Thomas Le Roux, Dr John Clark, Dr Robert Kirk, and Dr Roderick Bailey provided valuable comments and encouragement during the early stages of the project. Oxford University's 2013 conference on Rachel Carson and Ruth Harrison allowed me to establish contacts to people who knew Ruth Harrison. Dr Frank Uekötter encouraged me to continue looking for publishers, and a 2019 workshop by the Animal Research Nexus inspired me to integrate an analysis of the wider world of welfare research. The resulting manuscript has been immeasurably improved thanks to generous comments by Dr Dmitriy Myelnikov, Prof. Henry Buller, Prof. Donald Broom, and Dr Ashley Maher. My reviewers deserve double praise for their invaluable feedback and for rigorously reviewing a book amidst a global pandemic. All remaining mistakes are my own.

I am also grateful to the many organisations and individuals who granted me access to the written material and memories that lie at the heart of this study. The Whitechapel Gallery provided me access to Clare

Winsten's autobiography. Prof. Marian Stamp Dawkins agreed to be interviewed for this project, granted access to FACT's files, established contact with other FACT Trustees, and read drafts of this project. Prof. Donald Broom also agreed to be interviewed and allowed me to usurp his group's common room to pore over FACT's archives. Dr Ruth Layton provided valuable details about Ruth Harrison's later campaigning as well as Harrison's own copy of *Animal Machines*. I owe a special debt of gratitude to Phil Browning and the RSPCA for allowing me to access their archives—not once but twice because of a broken camera. I am also grateful to Prof. Peter Singer, Dr Richard Ryder, and Jonathan Harrison for answering queries about Ruth Harrison. The British Library provided access to Richard Ryder's papers, Yale's Beinecke Library to Ruth Harrison's correspondence with Rachel Carson, and the Library of the Society of Friends (London) to Harrison's Friends Ambulance Unit records, Cambridge University Library to William Homan Thorpe's papers, and the British National Archives to MAFF files on *Animal Machines*, the Brambell Committee, and FAWAC/FAWC. My research was financed by the Wellcome Trust and supported by the Oxford Martin School.

Finally, I would like to thank my family for their inexhaustible love and support. Charlotte, Clara, and Emil are more important to my research than they will ever appreciate. Countless dinner discussions with my wider family about animal welfare, consumer values, and expensive toys for pigs were also never far from my mind.

Abstract

Bearing Witness is the biography of one of Britain's foremost animal welfare campaigners and of the world of activism, science, and politics she inhabited. In 1964, Ruth Harrison's bestseller *Animal Machines* triggered a gear change in modern animal protection by popularising the term 'factory farming' alongside a new way of thinking about animal welfare. Here, historian Claas Kirchhelle explores Harrison's *avant-garde* upbringing, Quakerism, and how animal welfare debates were linked to concerns about the wider ethical and environmental trajectories of post-war Britain. Breaking the myth of Harrison as a one-hit wonder, Kirchhelle reconstructs Harrison's 46 years of campaigning and the rapid transformation of welfare politics and science during this time. Exacerbated by Harrison's own actions, the decades after 1964 saw a polarisation of animal politics, a professionalisation of British activism, and the rise of a new animal welfare science. Harrison's belief in incremental reform allowed her to form ties to leading scientists but alienated her from more radical campaigners. Many of her 1964 demands gradually became part of mainstream politics. However, farm animal welfare's increasing marketisation has also led to a relative divorce from the wider agenda of social improvement that Harrison once bore witness to.

Contents

1 Introduction 1

Part I Radical Roots (1920–1961) 19

2 Meet the Winstens: A 'Downstart' Anglo-Jewish family 21

3 Becoming an Activist: Ruth Harrison's Turn to Animal Welfare 35

Part II Synthesis: The Post-war Landscape of Welfare Science and Activism (1945–1964) 49

4 Between Physiology and Psychology—Ethology and Animal Feelings 51

5 Ideals and Intensification: Welfare Campaigns in a Nation of Animal Lovers 65

6 Staging Welfare: Writing *Animal Machines* 79

Part III Impact (1964–1968) 93

7 From Author to Adviser: Ruth Harrison and the *Animal Machines* Moment 95

Part IV Defining Welfare (1967–1979) 125

8 A "minority of one": Harrison and the FAWAC 127

9 Ruth the Ruthless: Activism, Welfare, and Generational Change 149

10 Slippery FACTs: The Rise of a "mandated" Animal Welfare Science 175

Part V From Éclat to Consensus (1979–2000) 203

11 From Protest to 'Holy Writ': The Mainstreaming of Welfare Politics 205

12 Non-conform Evidence: The Impasse of 1990s Welfare Research 223

13 Conclusion 239

Bibliography 249

Index 267

Abbreviations

ALF	Animal Liberation Front
BF	British Farmer
BFSS	British Field Sports Society
BVA	British Veterinary Association
CND	Campaign for Nuclear Disarmament
DB	Donald Broom Ruth Harrison Papers
EEC	European Economic Community
FACT	Farm Animal Care Trust
FAWAC	Farm Animal Welfare Advisory Committee
FAWC	Farm Animal Welfare Council
FW	Farmers Weekly
HSTA	Bayerisches Hauptstaatsarchiv [*Bavarian Main State Archive*]
LACS	League Against Cruel Sports
LSF	Library of the Society of Friends (London)
MAFF	Ministry of Agriculture Fisheries and Farms
MD	Marian Dawkins Ruth Harrison Papers
NFU	National Farmers' Union of England and Wales
OBE	Order of the British Empire
RADA	Royal Academy of Dramatic Art
RCVS	Royal College of Veterinary Surgeons
RSPCA	Royal Society for the Prevention of Cruelty to Animals
RVC	Royal Veterinary College
SVS	State Veterinary Service
T-AP	Standing Committee of the European Convention for the Protection of Animals Kept for Farming Purposes
TNA	The British National Archives

UFAW	Universities Federation for Animal Welfare
WFPA	World Federation for the Protection of Animals
WSPA	World Society for the Protection of Animals
YBL	Yale Beinecke Rare Book and Manuscript Library

List of Images

Image 2.1	Clare Winsten, Portrait by Isaac Rosenberg, oil on canvas 1916 (image courtesy of UCL Art Gallery and Bridgeman Images)	24
Image 2.2	Winsten Family Portrait of Clare, Theodora, Ruth, and Stephen, Photograph ca. 1922 (image courtesy of Jonathan Harrison)	30
Image 2.3	George Bernard Shaw by Clare Winsten, pencil on paper laid on board, 1949, NPG 6891 (image courtesy of National Portrait Gallery)	33
Image 3.1	Picture in Friends' Ambulance Unit Card Register (image courtesy of Friends' Library, London)	39
Image 3.2	Ruth Winsten Personnel File (image courtesy of Friends' Library, London)	40
Image 3.3	Theodora and Stephen Winsten, George Bernard Shaw, Ruth Winsten, Devdas Gandhi (son of Mohandas Gandhi) and Clare Winsten at Ayot St. Lawrence (ca. 1949) (image courtesy of Jonathan Harrison)	43
Image 3.4	Dex Harrison's pleasure gardens at the 1951 Festival of Britain (image courtesy of Jonathan Harrison)	44
Image 3.5	Leaflet by the Crusade Against All Cruelty To Animals pushed through Harrison's letterbox around 1960 (image courtesy of Marlene Halverson)	47
Image 6.1	Cover of *Animal Machines* (Ruth Harrison's personal copy) (image courtesy of Ruth Layton)	82
Image 6.2	*Animal Machines* image of a veal calf looking out of its crate with the neighbouring crate shut (image courtesy of Jonathan Harrison)	83

Image 11.1	Ruth Harrison and Klaus Vestergaard observe a sow at the Swedish Pig Park in 1988 (image courtesy of Bo Algers)	216
Image 11.2	Ruth Harrison at a Danish mink farm in 1997 (image courtesy of Marlene Halverson)	221

CHAPTER 1

Introduction

It is a rare event for a 40-year-old member of the public to pick up the mail and decide to write a book on farm animal welfare. Yet, supposedly, this is what Ruth Harrison (*née Winsten*) did in 1960. The result was an international bestseller and a turning point in the history of farm animal welfare. Often compared to American biologist Rachel Carson's environmentalist mile stone *Silent Spring* (1962), Harrison's *Animal Machines* (1964) has been cited as a major inspiration by animal welfare scientists around the world, and the regulatory impact of its publication has become part of the post-war activist lore.[1] The same cannot be said about the book's author. Now mostly remembered for her bestseller, Ruth Harrison often features as a one-hit-wonder, who suddenly emerged out of and then vanished back into the general ferment of British civic activism.[2]

[1] Ruth Harrison, *Animal Machines* (London: Vincent Stuart Ltd, 1964).

[2] Donald M. Broom, "A History of Animal Welfare Science," *Acta Biotheor* 59 (2011), 121–137; Donald M. Broom and Andrew F. Fraser. *Domestic animal behaviour and welfare* (Wallingford and Oxford: CABI, 2015); Linda J. Keeling, Jeff Rushen, and Ian JH Duncan. "Understanding animal welfare," in Michael C. Appleby, Anna Olsson, and Francisco Galindo (eds.), *Animal welfare* (Wallingford and Oxford: CABI, 2018), 13–26; Mieke Roscher, *Ein Königreich Für Tiere. Die Geschichte Der Britischen Tierrrechtsbewegung* (Marburg: Tectum Verlag, 2009), 260–261. Abigail Woods, "From Cruelty to Welfare: The Emergence of Farm Animal Welfare in Britain, 1964–71," *Endeavour* 36/1 (2012), 14–22; Karen Sayer, "Animal Machines: The Public Response to Intensification in Great Britain, C. 1960–C. 1973," *Agricultural History* 87/4 (2013), 473–501.

C. Kirchhelle, *Bearing Witness*, Palgrave Studies in the History of Social Movements,
https://doi.org/10.1007/978-3-030-62792-8_1

This narrative is too simple. Neither did the author Ruth Harrison appear by chance, nor did the campaigner Ruth Harrison vanish after 1964. A fuller analysis of Harrison's life instead reveals a remarkable campaigning career spanning much of the twentieth century and influencing the trajectory of British animal welfare politics, science, and activism until her death in 2000. It also reveals a complex world of different actor groups trying to come to terms with the changing contours of post-war Britain. This world comprised a who-is-who of leading intellectuals, campaigners, and decision-makers. It was also a place in which animal welfare was rarely just about animals. Although actors' concerns for animal welfare were genuine and deeply felt, they often symbolically stood in for broader concerns about the environmental and moral trajectories of British society.

Both concerns and proposed solutions changed considerably in the course of Ruth Harrison's life. While the influence of her vegetarian parents and Quaker beliefs loomed large over Harrison's own campaigning, the decades after 1945 saw many older forms of civic activism and thinking about animals' place in society change. Economically and intellectually, pre-war welfare arrangements were strained by a growing number of confined intensive animal husbandry operations—so-called factory farms—and by new concepts of humans' duties towards animals and animals' own rights. At the societal level, a younger generation of grassroots activists experimented with new forms of direct protest and challenged traditional animal politics and leadership structures. Scientifically, ethologists and veterinary researchers opened the door for new ways of conceptualising animal welfare in physical, behavioural, and cognitive terms. Politically, once powerful bastions of agricultural decision-making were complemented by a new set of non-governmental actors including professionalising welfare organisations, large retailers, and influential assurance schemes. By the end of the century, new animal welfare concepts had transformed the production, conceptualisation, and treatment of most British farm animals.

Bearing Witness is a biography both of Ruth Harrison and of the remarkable world of activism, scientific thinking, and politics she inhabited. The book follows core conventions of the literary genre of biography as defined by Hermione Lee: it starts with Harrison's birth in 1920 and ends with her death in 2000, tries to be as impartial as possible, investigates Harrison's identity, but also engages with biography as a form of history.[3] This latter point is particularly important. Inspired by the renaissance of life stories' approaches in the history of science and environmental, postcolonial, and legal studies,[4] the aim is not to provide a comprehensive inventory of events and actors. Nor is it to write a hagiography of Harrison. Instead, *Bearing Witness* follows recent research by Sally Sheldon, Gayle Davis, Jane O'Neill, and Clare Parker and uses Harrison's biography as an "important window onto the world around it."[5]

This approach has three advantages: (1) the chronology of Harrison's life parallels a significant period of agricultural change, which was associated with new forms of intensive food production and an increasing divorce of urban ideals of the rural from agricultural realities.[6] By tracing and contextualising Harrison's life, *Bearing Witness* is able to chart both long-term continuities and transition points in British thinking about farm animal welfare. (2) Harrison's hybrid status as a political insider and activist outsider also means that a biographic approach is uniquely suited to simultaneously examine what Angela Cassidy calls the public "frontstage" and compromise-oriented "backstage"[7] of welfare politics. Doing so over a longer period is all the more important given what Jon Agar calls the

[3] Hermione Lee, *Biography: A Very Short Introduction* (Oxford: Oxford University Press, 2009), 6–18.

[4] See, for example, Christoph Gradmann, *Laboratory Disease: Robert Koch's Medical Bacteriology*, Elborg Forster (trans.), (Baltimore: Johns Hopkins University Press, 2009); Adrian Desmond and James Moore, *Darwin's Sacred Cause: Race, Slavery and the Quest for Human Origins* (Chicago: University of Chicago Press, 2011); Sally Sheldon, Gayle Davis, Jane O'Neill, and Clare Parker, "The Abortion Act (1967): a biography," *Legal Studies* 39/1 (2019), 18–35; Mark Hamilton Lytle, *The gentle subversive: Rachel Carson, Silent Spring, and the rise of the environmental movement* (Oxford: Oxford University Press, 2007); David Nasaw, "AHR Roundtable: Historians and Biography," *American Historical Review* 114 (2009), 573; Judith M. Brown, "'Life Histories' and the History of Modern South Asia," *American Historical Review* 114/3 (2009), 587–595; Anna Lowenhaupt Tsing, *The mushroom at the end of the world: On the possibility of life in capitalist ruins* (Princeton: Princeton University Press, 2015); Sudipta Sen, *Ganges: The Many Pasts of an Indian River* (New Haven: Yale University Press, 2019).

[5] Sheldon, Davis, O'Neill, and Parker, "The Abortion Act," 32.

[6] Deborah Fitzgerald, *Every Farm a Factory. The Industrial Ideal in American Agriculture* (New Haven and London: Yale University Press, 2003); John Martin, *The Development of British Agriculture since 1931* (London et al.: Macmillan & St Martin's Press, 2000); Karen Sayer, *Farm Animals in Britain*, 1850–2001 (New York: Taylor & Francis, 2018).

[7] Angela Cassidy, *Vermin, Victims and Disease. British Debates over Bovine Tuberculosis and Badgers* (London: Palgrave Macmillan, 2019), 205.

post-1950s' "sea change"[8] of public attitudes towards expert authority and the rise of non-governmental research and assurance schemes. (3) Pursuing a biographic approach allows *Bearing Witness* to merge a macroscale analysis of twentieth-century politics and science with an actor-centred human-level perspective of how a leading campaigner experienced the transforming world around her.

The book achieves this merging of biography and macroscale analysis by drawing on a wealth of published and unpublished materials. These include oral history interviews and correspondence with leading animal welfare scientists and activists, archival documents from the British government and the Royal Society for the Prevention of Cruelty to Animals (RSPCA), the previously unseen personal papers of Ruth Harrison and her research charity, the personal papers of former RSPCA Chairman Richard Ryder, the papers of Rachel Carson, the unpublished autobiography of Harrison's mother, as well as contemporary scientific and media publications.

Readers will notice that the animals featured in this book are mostly talked about and do not 'talk' themselves. Over the past two decades, the rapidly growing field of animal studies has cast light on the "entangled"[9] relationships between humans and non-humans. The resulting body of research has flattened perceived differences between humans and animals and highlighted dynamic multi-species ecologies in the case of disease vectors, laboratory research practices, animal breeding, conservation practices, food production, and popular culture. It has also drawn attention to the role of built environments like farms, laboratories, cages, and housing systems in reflecting evolving moral economies of care and violence.[10]

[8] Jon Agar, *Science in the 20th Century and Beyond* (Cambridge: Polity, 2012), 403–432.

[9] Karen Barad, *Meeting the Universe Halfway: Quantum Physics and the Entanglement of Matter and Meaning* (Durham, NC: Duke University Press, 2007), 3–38.

[10] For a far from complete selection of work by historians and other disciplines, see Keith Thomas, *Man and the natural world: Changing attitudes in England 1500–1800* (London: Penguin UK, 1991), Hilda Kean, *Animal rights: Political and social change in Britain since 1800* (London: Reaktion Books, 1998); Angela NH Creager and William C. Jordan, eds. *The animal-human boundary: historical perspectives.* Vol. 2 (Cambridge MA: Harvard University Press, 2002); Robert G.W. Kirk, *Reliable animals, responsible scientists: constructing standard laboratory animals in Britain c. 1919–1976* (London: PhD Thesis University of London, 2005); Neil Pemberton and Mike Worboys, *Rabies in Britain. Dogs, Disease and Culture, 1830–2000* (London: Palgrave Macmillan, 2006); Donna J. Harraway, *When Species Meet* (Minneapolis: University of Minnesota Press, 2007); Henry Buller and Emma Roe. "Modifying and commodifying farm animal welfare: The economisation of layer chickens,"

Bearing Witness draws on the rich body of work resulting from the "animal turn"[11] but remains in the human realm. Focusing on Ruth Harrison and the socio-scientific development of farm animal welfare—which often happened away from farms—the book fills important scholarship gaps by (1) revealing the importance of farm animal welfare within the broad church of post-war activism, (2) providing a first comprehensive biography of Britain's most influential farm animal welfare activist, and (3) highlighting the synthesist ideological drivers of post-war animal welfare as well as the increasing power of European and non-governmental actors in British welfare politics.

At first glance, it seems remarkable that Harrison's life and campaigning have so far been neglected by historians. Two reasons emerge: the first has to do with disciplinary priorities. Mirroring the strength of environmental history in US academia, many accounts of post-war activism tend to neglect European campaigners like Harrison in favour of American figures like Rachel Carson, whose book *Silent Spring* (US 1962; UK 1963)

Journal of Rural Studies 33 (2014), 141–149; Jamie Lorimer, *Wildlife in the Anthropocene: conservation after nature* (Minneapolis: University of Minnesota Press, 2015); Michael Bressalier, Angela Cassidy, and Abigail Woods, "One Health in history," in J. Zinsstag et al. (eds.) *One Health: The Theory and Practice of Integrated Health Approaches* (Oxfordshire: CABI, 2015), 1–15; Kristian Bjørkdahl and Tone Druglitrø, eds. *Animal housing and human-animal relations: Politics, practices and infrastructures* (London: Routledge, 2016); Gail F. Davies, Beth J. Greenhough, Pru Hobson-West, Robert GW Kirk, Ken Applebee, Laura C. Bellingan, Manuel Berdoy et al. "Developing a collaborative agenda for humanities and social scientific research on laboratory animal science and welfare," *PLoS One* 11/7 (2016), e0158791. Henry Buller and Emma Roe. *Food and animal welfare* (London: Bloomsbury Publishing, 2018); Nicole C. Nelson, *Model behavior: Animal experiments, complexity, and the genetics of psychiatric disorders* (Chicago: University of Chicago Press, 2018); Rachel Mason Dentinger and Abigail Woods, "Introduction to Working Across Species," *History and Philosophy of the Life Sciences* 40/30 (2018), 1–11; Hilda Kean and Philip Howell (eds.), *The Routledge Companion to Animal-Human History* (London: Routledge, 2018); Robert GW Kirk, Neil Pemberton, and Tom Quick, "Being well together? promoting health and well-being through more than human collaboration and companionship," *Medical Humanities* 45/1 (2019), 75–81; Angela Cassidy, *Vermin, victims and disease: British debates over bovine tuberculosis and badgers* (London: Palgrave Macmillan, 2019); Gail Davies, Richard Gorman, Beth Greenhough, Pru Hobson-West, Robert G.W. Kirk, Dmitriy Myelnikov, Alexandra Palmer et al., "Animal research nexus: a new approach to the connections between science, health and animal welfare," *Medical Humanities* 46/4 (2020), 499–511.

[11] Erika Andersson Cederholm, Amelie Björck, Kristina Jennbert and Ann-Sofie Lönngren (eds.), *Exploring the Animal Turn. Human-Animal Relations in Science, Society and Culture* (Lund: Pufendorf Institute for Advanced Studies, 2014), 5.

is often credited with sparking modern environmentalism.[12] Even in Britain, farm animal welfare campaigning often fails to feature in accounts of other forms of contemporary civic activism like the peace movement or environmentalism. Over the last three decades, Meredith Veldman, Adam Lent, and Jodi Burkett have discussed the CND and *Silent Spring's* impact on British activism but have ignored Harrison and *Animal Machines.*[13] Such a focus not only neglects an important British campaigner but also contributes to an artificial separation of post-war protest movements into thematic blocks even though many participants marched for environmentalism, peace, and animal welfare alike.

In animal studies and the history of science, a prominent focus on laboratory and 'wild' animals or on famous ethologists and philosophers has similarly facilitated a comparative neglect of farm animals—and by extension Ruth Harrison.[14] Recent accounts like Angela Cassidy's excellent history of badgers and bovine tuberculosis or Michael Tichelar's account of English opposition to blood sports do much to link the worlds of animal, conservation, and environmentalist politics. However, they either do not or only briefly mention farm animal welfare activism and Harrison.[15]

[12] David Kinkela, *Ddt and the American Century: Global Health, Environmental Politics, and the Pesticide That Changed the World* (Chapel Hill: University of North Carolina, 2011); Garry Kroll, "The 'Silent Springs' of Rachel Carson: Mass Media and the Origins of Modern Environmentalism," *Public Understanding of Science* 10/4 (2001), 403–405; Joachim Radkau, *Die Ära Der Ökologie. Eine Weltgeschichte* (C.H. Beck: München, 2011), 118–123; Edmund Russell, *War and Nature: Fighting Humans and Insects with Chemicals from World War I to "Silent Spring"* (Cambridge: Cambridge University Press, 2001); Carson's views on animals have been explored by Marc Bekoff and Jan Nystrom, "The Other Side of Silence: Rachel Carson's Views of Animals," *Human Ecology Review* 11/2 (2004), 186–200.

[13] Meredith Veldman, *Fantasy, the Bomb and the Greening of Britain. Romantic Protest, 1945–1980* (Cambridge: Cambridge University Press, 1994); Adam Lent, *British Social Movements since 1945. Sex, Colour, Peace and Power* (Basingstoke and New York: Palgrave, 2001); Jodi Burkett, "The Campaign for Nuclear Disarmament and Changing Attitudes Towards the Earth in the Nuclear Age," *British Journal for the History of Science* 45/4 (2012), 625–39.

[14] Larry Carbone, *What animals want: expertise and advocacy in laboratory animal welfare policy* (Oxford: Oxford University Press, 2004); Richard W. Burkhardt Jr, *Patterns of behavior: Konrad Lorenz, Niko Tinbergen, and the founding of ethology* (Chicago: University of Chicago Press, 2005); Robert GW Kirk, "A brave new animal for a brave new world: The British Laboratory Animals Bureau and the constitution of international standards of laboratory animal production and use, circa 1947–1968," *Isis* 101/1 (2010), 62–94; Dmitriy Myelnikov, "Tinkering with genes and embryos: the multiple invention of transgenic mice c. 1980," *History and technology* 35/4 (2019), 425–452; Robert Garner and Yewande Okuleye, *The Oxford Group and the Emergence of Animal Rights* (Oxford: Oxford University Press, 2020).

[15] Angela Cassidy, *Vermin, Victims and Disease*, 129; Michael Tichelar, *The History of Opposition to Blood Sports in Twentieth Century England* (New York: Routledge, 2017), 11 & 28.

Intellectual histories of the period also tend to neglect Harrison and other adherents of traditional contractualist welfare models in favour of more radical utilitarian or animal rights thinkers like Peter Singer, Richard Ryder, and Tom Regan.[16] Although some accounts are beginning to challenge the described farm-field-laboratory divide in animal history,[17] a full survey of international (farm) animal welfare activism, politics, economics, and science remains an important desideratum.

The second reason for the relative neglect of Harrison is that the few accounts explicitly addressing her tend to focus on *Animal Machines* and its immediate impact rather than the book's author. This chronological flattening of Harrison's life can in part be explained by her self-proclaimed status as an uncompromising "loner,"[18] who often struggled to find allies in larger organisations and whose story fits uneasily into existing historiographies of activism and ethology.

In ethology, textbooks and articles covering the discipline's history usually stress the importance of *Animal Machines* for increasing the discipline's public recognition. However, most accounts quickly move on to the subsequent development of animal welfare science and do not dwell on Harrison's important family background or on her subsequent career as a campaigner and research sponsor.[19] In 2013, the University of Oxford

[16] Gary Francione, *Rain Without Thunder: The Ideology of the Animal Rights Movement* (Philadelphia: Temple University Press, [1996] 2007); Rod Preece, *Awe for the Tiger, Love for the Lamb: A Chronicle of Sensibility to Animals* (London and New York: Routledge, 2002); Peter Singer, *Animal Liberation: A Personal View. Writings on an Ethical Life* (London: Fourth Estate, 2001), 293–302; Gary L. Francione and Anna E. Charlton, "Animal rights," Linda Kalof (ed.), *The Oxford handbook of animal studies* (Oxford: Oxford University Press, 2017), 25–40; Garner, Robert and Yewande Okuleye, *The Oxford Group and The Emergence of Animal Rights* (Oxford: Oxford University Press, 2020).

[17] Frank Uekötter and Amir Zelinger, "Die Feinen Unterschiede. Die Tierschutzbewegung und die Gegenwart der Geschichte," in Herwig Grimm and Carola Otterstedt (eds.), *Das Tier an sich. Disziplinenübergreifende Perspektiven für neue Wege im wissenschaftsbasierten Tierschutz* (Göttingen: Vandenhoeck & Ruprecht, 2012), 119–134; Robert Kirk, "The Invention of the 'Stressed Animal' and the Development of a Science of Animal Welfare, 1947–86," in David Cantor and Edmund Ramsden (eds.), *Stress, Shock, and Adaptation in the Twentieth Century* (Woodbridge and Rochester: University of Rochester Press, 2014), 241–263; Robert GW Kirk and Edmund Ramsden. "Working across species down on the farm: Howard S. Liddell and the development of comparative psychopathology, c. 1923–1962," *History and philosophy of the life sciences* 40/1 (2018), 1–29.

[18] Ena Kendall, "Ruth and the Ruthless," *The Vegetarian*/New Series No. 43 (April) (1975), 2.

[19] Broom, "A History of Animal Welfare Science"; Broom and Fraser, *Domestic animal behaviour and welfare*; Keeling, Rushen, and Duncan, "Understanding animal welfare".

organised a conference to commemorate Rachel Carson and Ruth Harrison and launch a reprint of *Animal Machines.* The reprint's preface contains comments by leading contemporary welfare researchers. However, most have little to say about Harrison's non-literary roles as a government advisor, campaigner, and research funder.[20] Two notable exceptions to most ethological accounts' focus on 1964 are a 2008 essay by Heleen van de Weerd and Victoria Sandilands, which provides a cursory overview of important events in Harrison's life and career, and Edward Eadie's focus on Harrison's campaigning within the so-called Eurogroup for Animal Welfare from the 1980s onwards.[21]

Even specialist histories of British animal welfare and rights have tended to flatten Harrison's role. While Hilda Kean's magisterial *Animal Rights: Political and Social Change in Britain Since 1800* discusses *Silent Spring*, it does not mention *Animal Machines.*[22] In their work on protests against live animal exports and the rise of British green politics, Alun Howkins and Linda Merricks reference *Animal Machines* as a key text for a younger generation of 1970s' and 1980s' activists but do not focus on the book's genesis or author.[23] Similarly, former RSPCA chairman Richard Ryder's often-autobiographical history of British animal protection—*Animal Revolution*—devotes only a brief paragraph to Harrison despite his repeated clashes with her.[24] Both Robert Garner's and Mieke Roscher's important histories of British animal protection discuss *Animal Machines'* role in opening the way for the 1965 Brambell inquiry into animal welfare,

[20] Ruth Harrison, *Animal Machines—New Edition* (Wallingford and Boston: CABI, 2013).

[21] Heleen Van De Weerd and Victoria Sandilands, "Bringing the Issue of Animal Welfare to the Public: A Biography of Ruth Harrison (1920–2000)," *Applied Animal Behaviour Science* 113 (2008), 404–410; Edward N. Eadie, *Understanding Animal Welfare. An Integrated Approach* (Heidelberg et al.: Springer, 2012), 19–30; see also R.C. Newberry and Victoria Sandilands, "Pioneers of applied ethology," in *Animals and us: 50 years and more of applied ethology* (Wageningen: Wageningen Academic Publishers, 2016), 175–192.

[22] Hilda Kean, *Animal Rights. Political and Social Change in Britain Since 1800* (London: Reaktion Books, 1998).

[23] Alun Howkins and Linda Merricks, "'Dewy-Eyed Veal Calves'. Live Animal Exports and Middle-Class Opinion, 1980–1995," *The Agricultural History Review* 48/1 (2000), 85–103; Linda Merricks, "Green Politics: Animal Rights, Vegetarianism and Naturism," in David Morley & Kevin Robins (eds.), *British Cultural Studies* (Oxford: Oxford University Press, 2001), 435–436.

[24] Richard D. Ryder, *Animal Revolution. Changing Attitudes Towards Speciesism* (Oxford and New York: Berg, 2000).

but limit their subsequent analysis of Harrison to her role on the margins of the so-called Oxford Group of animal rights thinkers.[25]

Focusing in more detail on the changing nature of (farm) animal welfare during the 1960s, historians Abigail Woods and Robert Kirk highlight Harrison's importance in challenging agricultural equations of animal productivity ('thrift') with welfare and in creating popular pressure for a reformulation of official welfare definitions. However, events leading up to Harrison's attack on 'factory farming,' links to the nascent environmentalist movement, and Harrison's subsequent campaigning remain undiscussed.[26] In 2013, an important essay by Karen Sayer studied *Animal Machines'* impact in more detail. Sayer argues that *Animal Machines'* success was based on a pastoral romantisation of Britain's rural past. She also points to the fact that many husbandry systems did not resemble the 'factory farms' described by Harrison in 1964. However, Sayer does not provide further details on the genesis of *Animal Machines* or on Harrison's later work within government and animal campaigning organisations.[27]

By limiting our focus to the 1960s, we run danger of reducing Harrison's career to an individual act of romanticised protest. We also too readily accept the contemporary media's—and to a certain extent, Harrison's own—heroic narrative of a humble outsider, who within three years researched, wrote, and published a transformative international bestseller. This is a missed opportunity. Even a cursory glance at Harrison's life reveals a rich web of contacts with other leading campaigners and influential scientists. Their shared interest in animal cognition, emotions, and welfare and concerns about the moral status of post-war Britain would exert a powerful influence on the subsequent trajectory of animal science.[28] Similarly, ending an analysis of Harrison's campaigning in the 1960s is to ignore her role in the 1970s' polarisation of protest, the fraying of official decision-making, and the rise of a new era of European farm animal welfare politics and commercial assurance schemes.

[25] Roscher, *Ein Königreich Für Tiere* 260–261; Robert Garner, *Animals, Politics and Morality* (Manchester and New York: Manchester University Press, 1993), 108–110.

[26] Robert G.W. Kirk, "The Invention of the 'Stressed Animal'," 241–63; Woods, "From Cruelty to Welfare".

[27] Karen Sayer, "Animal Machines".

[28] Robert G.W. Kirk, "Science and humanity: national culture, scientific freedom and the limits of animal experiment in Britain and America, 1949–1966"—presented at the 2019 LSE/ Animal Research Nexus "National Cultures of Care, Animals and Science" workshop, *in preparation*.

The five parts of *Bearing Witness* are designed to both overcome the chronological flattening of Harrison and reinsert farm animal welfare into the wider history of post-war British activism, science, and politics. Part I reveals that Harrison was by no means a nobody but a well-connected and educated individual with a strong family tradition of civic activism. Harrison's parents were the painter Clara Birnberg and the author Samuel Weinstein (later Clare and Stephen Winsten). Both Samuel and Clara grew up in Eastern European Jewish families and were founding members of the so-called Whitechapel Boys, an important *avant-garde* group of artists from London's East End. The couple were committed socialists, pacifists, and vegetarians, who acted on their beliefs. After moving to rural Ayot St Lawrence around 1939, the Winstens became close friends of their neighbour and fellow vegetarian George Bernard Shaw. For the Winstens and the many Edwardian intellectuals, artists, and activists with whom they engaged, refusing to harm animals was part of a wider synthesist moral agenda of societal reform, which ultimately centred on improving human welfare and ethics.

Understanding this synthesist concept of welfare and socio-moral improvement is key to explaining the post-1945 turn towards animal welfare by activists and leading scientists like William Homan Thorpe and Julian Huxley. Having experienced the barbarity of the Second World War and steeped in synthesist Edwardian thinking about science, society, and morality, British campaigners and researchers saw animal welfare as part of a broader quest for moral reform. This was also true for Ruth Harrison. Born in 1920, the proximity to leading vegetarian, social, and peace activists during her youth left a profound mark on her. In 1939, she enrolled as an English major in Bedford College and made the significant decision to join the theist Society of Friends (Quakers). During the Second World War, she was first evacuated to Cambridge and then joined the Quaker-led Friends' Ambulance Unit (FAU). As the daughter of ethnic Jews, she then made the remarkable decision to aid displaced refugees in post-war Germany. After her return to the UK in 1946, Harrison enrolled in the Royal Academy of Dramatic Act and was coached by Shaw. However, despite winning awards, she did not pursue an artistic career after graduating in 1948 but instead joined an architectural firm. Following her 1954 marriage to the firm's partner, Dexter Harrison, Harrison settled into the seemingly quiet life of a Kensington housewife.

This quietude did not last long. Living in London, Harrison soon gained first-hand experience with the new non-violent protest of the

Campaign for Nuclear Disarmament (CND). She was not the only Quaker to do so. As historian Frank Zelko has shown, Quakers' tradition of peacefully 'bearing witness' against unethical activities made them particularly active in the peace movement and nascent environmentalist groups after 1945.[29] Harrison provides a perfect example of Zelko's profile of mid-century Quaker environmentalists like Greenpeace founders Irving and Harriet Beecher Stowe: born into a highly educated Jewish family with influential cultural contacts, Harrison had already shown a remarkable commitment to the Quaker principle of living faith through action and 'bearing witness' during the war. Generationally, she stood between two key cohorts of British animal activism. Old enough to meet leading Edwardian figures like George Bernard Shaw, Harrison was also young enough to participate in and appreciate the power of new post-war protest movements like the CND. At the same time, her parents' political background and her economically constrained upbringing among Britain's elite enabled her to fluently converse with radical and establishment circles alike.

As Part II shows, Harrison's ability to move between older and younger as well as between establishment and anti-establishment circles made her perfectly poised to shape a watershed moment in animal welfare history. Beginning work on *Animal Machines* in 1961, Harrison was writing during a time of heightened wariness about the social, moral, and environmental side-effects of technological 'progress' as well as intensifying scientific engagement with animal cognition and emotions.

In the case of the animal sciences, the 1950s saw previously dominant behaviourist concepts of machine-like mental conditioning come under fire. Since around 1930, researchers belonging to the young discipline of ethology had begun to redirect attention to animals' evolutionarily acquired behaviours, ability to learn via insight, and complex social lives. Trying to avoid accusations of anthropomorphism, leading continental ethologists had, however, shied away from publicly engaging with charged debates about animal cognition in the context of animal welfare. Their British colleagues felt less compunction. Inspired by what Robert Kirk has called "scientific humanism,"[30] members of the Universities Federation for

[29] Frank Zelko, *Make It a Green Peace! The Rise of Countercultural Environmentalism* (Oxford: Oxford University Press, 2013).

[30] Robert G.W. Kirk, "Science and humanity: national culture, scientific freedom and the limits of animal experiment in Britain and America, 1949–1966," *in preparation*.

Animal Welfare (UFAW) looked for ways to scientifically improve animal welfare—and human society—whilst distancing themselves from 'anti-scientific' antivivisectionism. From the 1940s, UFAW researchers focused on developing quantifiable measures of stress as a way to make ethical concerns scientifically and politically reputable in the context of standardising laboratory animals' genetic, behavioural, and physiological traits. If science was a humane force for improving society, its methods had to be humane, too. In 1959, the UFAW publication *Principles of Humane Research* used this approach to lay out an enduring new agenda for animal laboratory research based on the principles of replacement, reduction, and refinement.[31]

Other prominent British researchers also focused on animal cognition and welfare as part of a wider quest to reconcile scientific and moral values. After 1945, British ethologist Julian Sorrell Huxley linked the humane treatment of animals to his wider vision of transhumanist social evolution. In Cambridge, ethologist William Homan Thorpe became interested in animal welfare as a result of his quest to reconcile Darwinian evolution with Christian salvation. Key to Thorpe's work was the concept of emergence—an evolutionary event where the outcome is greater than the sum of its parts. By stipulating that consciousness was a 'creative' emergent event, Thorpe could simultaneously argue that humans had descended from animals via non-random evolution but were also distinct and thus subject to Christian salvation.[32] Thorpe's subsequent research on animal behaviour and (insight) learning made important contributions to ethology and buttressed calls for positive definitions of welfare that were not limited to reducing cruelty.

The 'affective turn' allowed ethologists and affiliated scientists to present themselves as best-placed to answer resulting calls for new welfare standards. Despite parallel controversies about field sports (hunting for pleasure) and animal experimentation, the treatment of farm animals on new intensive animal production facilities dominated ensuing public controversies. Similar to nuclear energy, the 'factory farm' functioned as

[31] William Moy Stratton Russell and Rex Leonard Burch, *The principles of humane experimental technique* (London: Methuen, 1959); Robert G.W. Kirk, "Recovering the principles of humane experimental technique: The 3Rs and the human essence of animal research," *Science, Technology, & Human Values* 43/4 (2018), 622–648.

[32] Gregory Radick, "Animal agency in the age of the Modern Synthesis: W.H. Thorpe's example," *British Journal of the History of Science* Themes 2 (2017), 35–56; Neal C. Gillespie, "The Interface of Natural Theology and Science in the Ethology of W. H. Thorpe," *Journal of the History of Biology* 23/1 (1990), 1–38.

what Sheila Jasanoff and Sang-Hyun Kim have described as a dystopian 'sociotechnical imaginary'[33] and fused diverse strands of contemporary concern.

The dystopia of dehumanising farms had started as a utopia of agricultural plenty.[34] However, by 1960, the public image of still far from ubiquitous 'factory farms' was becoming ambivalent.[35] Proponents continued to present 'factory farms' as a progressive way to ward off overpopulation-induced famine. However, critics increasingly interpreted them as symbols of humans' industrial and scientific alienation from 'nature.' This alienation was presented as both physically and morally damaging. A series of contemporary bestsellers like William Longgood's *The Poisons in Your Food* (1960), Frances Bicknell's *Chemicals in Food and in Farm Produce* (1960), Doris Grant's *Your Bread Your Health* (1961), and Rachel Carson's *Silent Spring* (1962)[36] all stressed the physical and moral dangers of new farming practices and chemical technologies. Public fears were heightened by a series of health scares, intensified media reporting on food laden with 'chemical' or radioactive residues, and new data on the selection for antibiotic-resistant organisms on farms.[37] In addition to health fears, a second powerful strand of criticism centred on new production methods' alleged cruelty. Drawing on wartime tropes of Britain as a "nation of animal lovers,"[38] activists used emerging research on animals' affective states to accuse 'alien' confinement systems of jeopardising animals' emotional welfare and undermining British civility.

Ruth Harrison's ability to weave together these diverse environmental, moral, and welfare concerns about intensive food production was key to the success of *Animal Machines.* Invoking a romanticised pastoral past while tapping into new scientific concepts of cognition, Harrison produced a compelling dystopian imaginary centring on the cruel,

[33] Sheila Jasanoff and Sang-Hyun Kim, "Sociotechnical Imaginaries and National Energy Policies," *Science as Culture* 22/2 (2013), 189–196.

[34] Deborah Fitzgerald, *Every Farm a Factory: The Industrial Ideal in American Agriculture* (New Haven: Yale University Press, 2010).

[35] Sayer, "Animal Machines," 473–501.

[36] William Longgood, *The Poisons in Your Food* (New York: Simon and Schuster, 1960); Franklin Bicknell, *Chemicals in Food and in Farm Produce: Their Harmful Effects* (London: Faber and Faber, 1960); Doris Grant, *Your Bread and Your Life* (London: Faber and Faber, 1961); Rachel Carson, *Silent Spring* (Boston: Houghton Mifflin, 1962).

[37] Claas Kirchhelle, *Pyrrhic Progress: The History of Antibiotics in Anglo-American Food Production* (New Brunswick: Rutgers University Press, 2020), 17–32; 77–91.

[38] Kean, *Animal Rights,* 165–179, 191–200.

dehumanising, and unhealthy 'factory farm.' The dystopia was linked to a call for a redefinition of welfare that went beyond productivity and the absence of cruelty and encompassed physical and affective states.

Part III reconstructs how the 1964 publication of *Animal Machines* became a watershed moment in the history of animal welfare. Taking many by surprise, the book's societal impact was aided by Harrison's carefully constructed public image as a concerned citizen. Similar to what Emily Gaarder and Angela Cassidy have described for other female animal activists, Harrison's gender led to attempts by a predominantly male agricultural and veterinary establishment to downplay her concerns as 'overemotional,' anthropomorphic, and thus unscientific. However, Harrison's status as a charismatic outsider also enabled her to present herself as a trustworthy intermediary in a rapidly evolving campaigning field.[39] Skilfully occupying a middle ground between establishment campaigning groups like the RSPCA and more radical protestors, Harrison was able to mobilise sustained support for systemic animal welfare reform not just in activist but in wider public circles. Although she was not a member of the UK's Brambell Committee on animal welfare, the agenda set out in *Animal Machines* shaped both the establishment of the committee and its resulting 1965 report. Written by William Thorpe, the report's influential appendix on essential animal freedoms challenged the narrow models of welfare as the absence of pain and welfare as thrift that dominated up to that point.[40] Significantly, the committee's call for legislative reform and a standing committee on welfare also created new places at the policy table for outsiders like Harrison and behaviour-focused researchers—some of whom began calling themselves animal welfare scientists.

Part IV examines the resulting 1970s' explosion of regulatory, activist, and scientific welfare work. It shows how the expanding political arena of farm animal welfare triggered clashes over proposed welfare codes, a restructuring and professionalisation of campaigning groups like the RSPCA, and the rise of animal welfare science as a "mandated"[41] discipline tasked with providing welfare standards. For Ruth Harrison, her

[39] Cassidy, *Vermin, Victims, and Disease*, 183–186; Emily Gaarder, *Women and the Animal Rights Movement* (New Brunswick: Rutgers University Press, 2011).

[40] Abigail Woods, "From Cruelty to Welfare: The Emergence of Farm Animal Welfare in Britain, 1964–71," *Endeavour* 36/1 (2012), 14–22.

[41] David Fraser, "Understanding Animal Welfare," *Acta Veterinaria Scandinavica* 50/Supplement (2008), 1.

post-1960s' role as a lynchpin connecting regulators, scientists, and moderate and radical campaigners was often uneasy. At the regulatory level, her membership within the government's new Farm Animal Welfare Advisory Committee (FAWAC) and unwavering commitment to animal welfare often led her to act as a 'minority of one.' Employing a mixed strategy of targeted leaks, public campaigning, and specially commissioned research, dissent by Harrison and a small group of allies prevented FAWAC from 'rubberstamping' industry-friendly standards but also paralysed traditional corporatist decision-making.

Outside government circles, Harrison's position was even more complicated due to her paradoxical status as an establishment and anti-establishment figure. Harrison's often single-minded opposition to weak welfare provisions led to a notorious fallout with 'traditionalists' in the RSPCA Council over 'field sports' and Harrison's personal bankruptcy in 1975. However, despite her apparent radicalism, younger—often male—campaigners accused Harrison of being too 'timid' in pressing for the reform of rather than the abolition of intensive agriculture. Intergenerational disagreement encompassed both tactics and wider moral visions of society. Many older 'welfarist' campaigners like Harrison insisted on a contractarian notion of animal welfare on the grounds that animals had cognition, were fellow creatures of God, and that cruelty desensitised society. However, most maintained that there was a distinction between animal and human life. By contrast, more radical 1970s' thinkers like Richard Ryder, Tom Regan, and Peter Singer argued that the interests of humans and non-humans deserved equal consideration. Employing utilitarian reasoning or arguing from the standpoint that animals enjoyed inalienable rights, these younger activists opposed the 'speciesist' exploitation of animals for food, leisure, or science.[42]

Growing controversies about animal ethics created problems for official welfare committees and for animal protection organisations. In the case of the RSPCA, the early 1970s had seen a rapid professionalisation and expansion of RSPCA welfare lobbying and the sponsorship of targeted welfare research. However, the Society's parallel failure to clearly oppose elite 'field sports' and its exclusive leadership structures caused growing

[42] Henry Buller and Emma Roe, *Food and Animal Welfare* (London et al.: Bloomsbury Academic, 2018), 21–23; Clare Palmer and Peter Sandøe, "Animal ethics," in Michael C. Appleby, I. Anna S. Olsson, Francisco Galinda (eds.), *Animal welfare* (Oxford and Wallingford: CABI, 2018), 3–15; Garner and Okuleye, *The Oxford Group and The Emergence of Animal Rights*.

grassroots discontent. In 1975, a highly critical internal review triggered a significant shift of RSPCA campaigning and management. Led by members of the so-called RSPCA Reform Group, the second half of the 1970s saw the Society streamline its management, organise conferences on animal rights, and launch sophisticated and expensive campaigns against live animal exports. The Society's marked shift away from compromise-oriented lobbying to public campaigning and its 1979 boycott of the government's newly created Farm Animal Welfare Council (FAWC) divided RSPCA members. It also complicated the Society's formerly close relations with animal scientists, whose findings were becoming increasingly important for campaigning but whose methods contradicted more radical Reform Group members' opposition to animal exploitation.

Welfare researchers themselves had to strike a balance between maintaining scientific authority and producing findings that had 'practical value' to their official, industrial, and activist sponsors like Ruth Harrison and the RSPCA. While rising funding and the ongoing demand for welfare standards attracted talented researchers, there was still no agreement on how to define, measure, and interpret seemingly basic parameters such as stress, 'natural' and 'abnormal' behaviour, or animal 'feelings.' Fifteen years after *Animal Machines*, British animal welfare politics had seemingly become bogged down by scientific uncertainty, ethical disagreements, and a breakdown of consultative official and activist decision-making.

Part V explores how the described crisis created an opening for new forms of market-driven welfare politics and strengthened moderate activists like Harrison. While 1979 marked a highpoint of discontent and stasis, the two subsequent decades saw a weakening of traditional bastions of pro-industry corporatist decision-making as well as of more radical activists. With animal welfare demands becoming part of mainstream politics, new actor coalitions emerged. These coalitions spanned governmental and non-governmental circles. Although official British and European welfare bodies remained important, private assurance schemes for animal welfare became powerful drivers of animal welfare politics. Formed in response to consumer demand and market segmentation, welfare certification schemes created lucrative alliances between major retailers, animal welfare organisations like the RSPCA, and other non-official bodies tasked with monitoring private standards. Despite ongoing controversies about whether it was possible to establish universal welfare standards, the continuous growth of well-financed welfare schemes also benefited welfare researchers. From around 1980 onwards, the number of welfare-related publications began to soar. Prominent welfare researchers were also appointed to important university posts and managed to expand their influence in academic and official circles. The same was true for Ruth Harrison. Having survived the tumultuous 1970s, Harrison retained her membership on important British and European welfare committees, cultivated relations with leading researchers, and witnessed the fulfilment of core demands of *Animal Machines* like a ban of veal crates. After continuously campaigning for animal welfare for four decades, Harrison died in 2000. By this time, many of the values she had 'born witness' to in 1964 had become firmly entrenched in all spheres of British welfare politics, activism, and science.

PART I

Radical Roots (1920–1961)

The bestselling author Ruth Harrison did not emerge by chance. While the established narrative of a concerned citizen, who spontaneously wrote *Animal Machines*, is attractive, Harrison's success as a campaigner is far less surprising—though no less remarkable—when one studies her life prior to 1960. As Part I shows, Harrison grew up surrounded by radical Edwardian intellectuals, vegetarians, and pacifists. Although they never achieved the fame of friends like Isaac Rosenberg, her parents Clara Birnbaum and Samuel Weinstein (later Clare and Stephen Winsten) were active members of the Anglo-Jewish Whitechapel Boys and skilfully cultivated connections throughout Britain's cultural establishment. The Winstens' upward social mobility left a mark on their children, who became successful academics and artists themselves. In the case of Ruth, her parents' pacifist, vegetarian, and activist values provide an important context for her decision to convert to Quakerism around 1939. They also laid the foundations for Harrison's synthesist worldview of social and moral improvement that would motivate her to alleviate human suffering during the Second World War, protest against 1950s' nuclear armament, and later bear witness to farm animals' perceived plight. Well-educated, experienced in campaigning, and steeped in synthesist Edwardian values, Harrison was perfectly positioned to campaign for animal welfare reform against the dystopian backdrop of the factory farm.

CHAPTER 2

Meet the Winstens: A 'Downstart' Anglo-Jewish family

Ruth Harrison was born as Ruth Winsten on June 24, 1920,[1] into a highly intellectual and artistic household. Her parents' immigrant roots and outspoken commitment to pacifism, vegetarianism, and socialist welfare would have a profound impact on Harrison's upbringing and later campaigning. Despite her intimate contact to leading Edwardian radicals, artists, and intellectuals, Ruth's childhood was also characterised by persistent economic insecurity. The sum total of these experiences was a campaigner inculcated with the synthesist ideals of Edwardian reform, capable of moving confidently amongst Britain's upper and middle classes, but equally comfortable with sacrificing economic well-being for moral victories. For the Winstens, the treatment of animals was always part of a wider social and ethical reform agenda.

Before the First World War, Harrison's father, Samuel–'Simy'/'Sammy'/'Simon'–Weinstein (1893–1991), and her mother, Clara Birnberg (ca. 1892–1989), belonged to a group of *avant-garde* writers and artists from London's East End. Retrospectively known as the 'Whitechapel Boys,' the group had close ties with the influential Slade School of Art and met regularly to discuss politics, art, and literature. Among its members were Isaac Rosenberg, John Rodker, Joseph Leftwich, David Bomberg, and Mark Gertler. Jacob Epstein and Sonia Cohen also

[1] Richard D. Ryder, "Harrison, Ruth (1920–2000)," *Oxford Dictionary of National Biography* (Oxford University Press, 2004).

C. Kirchhelle, *Bearing Witness*, Palgrave Studies in the History of Social Movements,
https://doi.org/10.1007/978-3-030-62792-8_2

had contact with the group. In contrast to the affluent and nearly contemporaneous Bloomsbury group, members of the 'Whitechapel Boys' came from mostly poor, Eastern European Jewish backgrounds.[2]

Clara Birnberg was a painter whose work was included in the 1914 Whitechapel exhibition on *Twentieth Century Art: A Review of Modern Movements*.[3] In 1915, Isaac Rosenberg portrayed her as *Girl in a Red Dress*.[4] David Bomberg also painted her as *Sibyl* in an earlier, now-lost painting. Birnberg had been born in Romania in 1892 as the second daughter of Michael, a former teacher, and Fanny Birnberg. The Birnbergs were originally from Tarnopol in Galicia but had been forced to flee as a result of pogroms from what is now modern Ukraine to Romania. The family subsequently moved to London where Michael had been promised employment, which failed to materialise. Highly educated but speaking limited English, the family instead established a small and not very profitable cabinet-making business on Leman Street in Aldgate. According to Clara Birnberg's unpublished autobiography, life was hard. Arriving in London to join Michael on Edward VII's coronation day in 1902, Clara was struck by the crass inequality of Edwardian Britain:

> we found ourselves stranded on the docks among large boxes and sacks full of grain. We expected my father to await us but by some error of time he did not come (…). Soon all [waiting passengers] were bundled into a van covered with canvas and drawn by two horses. (…). We had the opportunity to see the unswept streets, the horror of men and women lounging along the sordid pavements, children lying about (…), and amidst all drunken men and women dancing and singing: "On the coronation day" (…) My heart sank as I saw the shabby little houses (…). This, our first day in London has ever stayed in my mind. As though awakened from a deep sleep to find oneself in hell.[5]

Economic circumstances were difficult. Later describing herself as a 'downstart,' Clara remembered debt collectors calling at the family's

[2] Rachel Dickson and Sarah Macdougall, "The Whitechapel Boys," *Jewish Quarterly* 51/3 (2004), 29–30; 32–33, Jean Moorcroft Wilson, *Isaac Rosenberg: The Making of a Great War Poet: A New Life* 2nd edition (Chicago: Northwestern University Press, 2009), 94–95.

[3] Dickson and Macdougall, "The Whitechapel Boys," 29 & 34.

[4] Wilson, *Isaac Rosenberg*, 101.

[5] Whitechapel Gallery Archive, Clare Winsten Autobiography, WG/DON/1, [subsequently, Winsten Autobiography; I have used the page numbers on the typed manuscript despite a doubling of page numbers in the original manuscript], 25–27; see also 46–47, 50.

house and being forced to continuously economise at home and school. Despite their relative poverty, the family of five still managed to lead a rich cultural life. Living in immediate proximity to the recently built Whitechapel Gallery and Toynbee Hall, the Birnbergs quickly became part of the vibrant artistic, musical, and intellectual milieu of London's East End. Michael Birnberg was later described as one of the new Anglo-Jewish intellectuals of Whitechapel, and Clara and her younger brother, Jonas (or 'Johanes,' b. 1894 in Galatz, died 1970), were able to win prestigious scholarships from London's County Council.[6]

Their academic prowess enabled both children to climb Britain's social ladder. Jonas obtained a stipend to study mathematics at Queen's College Cambridge, was London's chess champion in 1924 and 1935, and taught at Corfe Grammar School and Goldsmiths College (now University of London). In 1928, he married his Cambridge contemporary Naomi Bentwich (born 1891), daughter of Herbert Bentwich, founder of the Hampstead Synagogue, ex-secretary to John Maynard Keynes, and founder of the vegetarian Carmelcourt School.[7] Clara's 'downstart' career was also rapid. After first visiting Rutland Street Council School, she was soon admitted to Central Foundation Girls School (Tower Hamlets). Her artistic talent enabled Clara to win a scholarship to the Royal Female School of Art, which was part of the Central School of Art and Design (today Central St Martins) in 1910.[8] Shortly afterwards, on the "strength of her promise" Clara transferred to the Slade School of Fine Art, where she studied between 1910 and 1912.[9] Although Jonas was critical of Clara's commitment to a potentially penniless artistic career, her parents used their meagre resources to help Clara rent her own studio on City Road.[10]

[6] Winsten Autobiography, 46–47; Sarah Macdougall, "Whitechapel Girl: Clare Winsten and Isaac Rosenberg," in Sarah Macdougall, Dickson Rachel, and Ben Uri Art Gallery (eds.), *Whitechapel at War: Isaac Rosenberg & His Circle* (London: Ben Uri Gallery 2008), 100.

[7] "Birnberg, Benedict Michael," in W. Rubinstein and Michael Jolles (eds), *The Palgrave Dictionary of Anglo-Jewish History*; Ariadne Birnberg, *Most Beautiful Maynard*, https://longandvariable.files.wordpress.com/2015/05/most-beautiful-maynard.pdf [01.05.2020]; Winsten Autobiography, 44.

[8] Macdougall, "Whitechapel Girl," 100.

[9] Sarah Macdougall, "'Something Is Happening There': Early British Modernism, the Great War and the 'Whitechapel Boys'," in Michael J. K. Walsh (ed.), *London, Modernism, and 1914* (Cambridge: Cambridge University Press, 2010), 127, the Slade School offers two annual Clare Winsten Memorial Award for its female students.

[10] Winsten Autobiography, 44–45, 52A; MacDougall, "Whitechapel Girl," 99.

At Slade, Clara maintained a low profile as the only female working-class Jewish student of her generation. However, outside of Slade, the early 1910s saw her gain increasing visibility as the sole "Girl" member of the Whitechapel Boys.[11] Involved in a stormy relationship with David Bomberg, Birnberg's early paintings (e.g. *Dawn* [ca. 1912]) were influenced by Post-Impressionism and Vorticism.[12] By 1913, she had also enrolled in book illustration and sculpture courses at the Central School and Slade to be "free of financial worries while practicing as a painter."[13] Her ability to work and generate income in multiple media as well as her Slade connections would turn out to be a vital asset (Image 2.1).

Image 2.1 Clare Winsten, Portrait by Isaac Rosenberg, oil on canvas 1916 (image courtesy of UCL Art Gallery and Bridgeman Images)

[11] Macdougall, "Whitechapel Girl," 99 & 102.

[12] Macdougall, "Something Is Happening There," 131–32, Macdougall, "Whitechapel Girl," 99–108.

[13] Winsten Autobiography, 53 & 55.

Throughout this period, the entire Birnberg family engaged actively in contemporary politics. In her autobiography, Clara Birnberg recalls becoming politically sensitised to social injustice and women's rights while attending school in London's East End. Together with a friend, she would regularly go "to the meetings in Hyde Park or the Embankment Gardens" and "offered to sell [suffragette publications] in different places."[14] Doing so was not without risks. Clara describes having to avoid intimate approaches by young men in the crowds, police violence, as well as being assaulted by an old man while selling *The Freedom League*. When suffragette Emily Davison was trampled to death by King George V's horse in 1913, all three Birnberg women joined the funeral march. Another significant event for the family was Clara's conversion to vegetarianism around 1910. Trying to take a short cut through a small passage one morning, Clara:

> saw a group of little boys staring eagerly into a crack of a door, leading into a wooden building. A man had also joined them. Suddenly I heard a scuffle, men cursing gruffly, piteous cries of shuffling cattle … a pool of blood streaming out from underneath the wooden doors. I rushed madly out, realising furiously that there was murder behind those doors, hating the eager sense of curiosity shewn by the man and the boys, relishing slaughter! (…) That evening when I came home I told my mother and my sister that I would never eat meat.[15]

Her entire family converted to vegetarianism alongside her.

As described by historian Elsa Richardson, converting to vegetarianism was no trivial choice and carried specific political connotations. Since 1847, Britain's Vegetarian Society had drawn support from a mix of working- and middle-class campaigners and religious dissenters. Although it is not clear whether the Birnbergs formally joined the Society, their personal politics aligned closely with the vegetarian movement's progressivist advocacy for a wider moral and welfarist reform of society. At the heart of this advocacy was a contractarian notion of universal kinship between humans and animals. This kinship entailed moral responsibility for the welfare of all animal and human life and a condemnation of 'flesh eating' as a form of spiritual and physical desecration. Drawing heavily on evolutionary theories, key campaigners argued that reforming one's own diet was a

[14] Winsten Autobiography, 37.

[15] Winsten Autobiography, 43.

necessary component of creating a more progressive, cooperative, and egalitarian society.[16]

Clara's synthesist ethics of universal kinship and social and spiritual improvement were shared by her fiancé and fellow Whitechapel Boy Samuel Weinstein. 'Simy' had been part of the Whitechapel Boys[17] from the beginning and had grown up only a few streets away from Clara. Preparing to go to teacher-training college after a brief stint at the London School of Economics,[18] Weinstein came from a Russian-Jewish family with strong Marxist sympathies. His father was a "bearded Jewish scholar-type"[19] from Russian Vilna. His older brother Rachmiel (Aaron) had stayed in Russia, where he became a labour leader and prominent figure within the Jewish social democratic Bund, which had been founded in Vilna in 1897. Temporarily deported by the Tsarist regime for resisting conscription,[20] Rachmiel joined the subsequent revolution and was appointed Soviet Commissioner and member of the Committee for Settling Jews on the Land in Ukraine in 1924 before falling victim to Stalinist purges under Beria.[21] Another of Simy's siblings, Mary, was an outspoken London Zionist, and her husband Zalkind Stalbow was active in *Bnai Zion*.[22]

Simy shared his family's internationalist leanings and was also a convinced pacifist. In 1911, he joined the Young Socialists together with Isaac Rosenberg.[23] Working as a teacher in an East End Board school, he taught one of Rosenberg's younger brothers and was active in numerous local societies like the Ben Uri Art Society and the Jewish Association for Advancement in Arts and Sciences.[24] Following the outbreak of the First

[16] Elsa Richardson, "Man is not a meat-eating animal: vegetarians and evolution in late-Victorian Britain," *Victorian Review* 45/1 (2019), 117–134; see also: James Gregory, *Of Victorians and Vegetarians: The Vegetarian Movement in Victorian Britain* (London and New York: Tauris, 2007).

[17] Dickson and Macdougall, "The Whitechapel Boys," 30 & 34.

[18] Macdougall, "Whitechapel Girl," 110; Winsten Autobiography, 59–60.

[19] Quoted according to Macdougall, "Whitechapel Girl," 109.

[20] "Stephen Winsten, 1893–1991", *Remembering the men who said no, conscientious objectors 1916–1919, Peace Pledge Union project*, https://menwhosaidno.org/men/men_files/w/winstent_s.html [30.04.2020].

[21] "Soviet Government Will Not Interfere with Administration of Funds for Jewish Colonization Work in Russia," *Jewish Telegraph*, 06.10.1924; "Communist Paper Publishes List of Executed Soviet Jewish Intellectuals," *Jewish Telegraph*, 12.04.1956; Macdougall, "Whitechapel Girl," 109.

[22] Macdougall, "Whitechapel Girl," 109, Wilson, *Isaac Rosenberg*, 101.

[23] Wilson, *Isaac Rosenberg*, 101.

[24] Macdougall, "Something Is Happening There," 126; 134; Dickson and MacDougall, "The Whitechapel Boys," 34.

World War in 1914, he joined the pacifist No Conscription Fellowship and the Peace Pledge Union alongside his fiancé, Clara Birnberg, and their mutual friends Isaac Rosenberg, and—the later promoter of Soviet art and literature—John Rodker.[25] While Rosenberg was forced to enlist in 1915 for financial reasons, Weinstein, Rodker, and Clara's brother Jonas resisted the introduction of conscription in Britain in 1916.[26]

To prepare for his upcoming trial, Simy Weinstein enlisted the help of prominent anti-war activist and Labour politician Fenner Brockway, who advised him to contact Tolstoy's biographer Aylmer Maude for a witness statement.[27] According to historian Ann Kramer, Maude warned officials that pacifism came natural to Weinstein, who would be a "nuisance"[28] if forced to fight. This strategy backfired. When Simy formally registered his objection to conscription at the Hackney Tribunal where he was working as a supply teacher in 1916, his claim that he knew "what it is to kill a pig, I will not kill a man"[29] fell on death ears. Tribunal members asked whether the fact that Britain had provided him with a teaching job did not mean that he owed "something to the country."[30] According to Weinstein:

> [The official] meant that I was a dirty cad… I was then teaching in the roughest school in London … so I said to him, 'It is because I *love* England sir, that I'm willing to serve in any position and do a service which I don't think you would ever do.' 'Well,' he said, 'I think we'll put you down as a political objector and therefore you can't get exemption. We can only give it to religious [objectors].[31]

[25] Wilson, *Isaac Rosenberg*, 142–43, Macdougall, "Something Is Happening There," 131–33, Macdougall, "Whitechapel Girl," 110–14; "Stephen Winsten, 1893–1991", *Remembering the men who said no, conscientious objectors 1916–1919, Peace Pledge Union project*, https://menwhosaidno.org/men/men_files/w/winstent_s.html [30.04.2020].

[26] Ian Patterson, "The Translation of Soviet Literature," in Rebecca Beasley and Philip Ross Bullock (eds.), *Russia in Britain, 1880–1940: From Melodrama to Modernism* (Oxford: Oxford University Press, 2013), 189–90, Macdougall, "Whitechapel Girl," 112.

[27] Ann Kramer, *Conscientious Objectors of the First World War: A Determined Resistance* (Barnsley: Pen & Sword, 2014), 53.

[28] Quoted according to Kramer, *Conscientious Objectors*, 53.

[29] Quoted according to "Stephen Winsten, 1893–1991", *Remembering the men who said no, conscientious objectors 1916–1919, Peace Pledge Union project*, https://menwhosaidno.org/men/men_files/w/winstent_s.html [30.04.2020]; other sources attribute this statement to a butcher whom Winsten met while he was imprisoned: Sara Ayad, "The Winstens of Whitechapel: Clara Birnberg and Simy Weinstein," Art UK, https://artuk.org/discover/stories/the-winstens-of-whitechapel-clara-birnberg-and-simy-weinstein [22.02.2021].

[30] Quoted according to Kramer, *Conscientious Objectors*, 53.

[31] Quoted according to Kramer, *Conscientious Objectors*, 53.

Perhaps hoping that prison would change his mind, the tribunal sentenced Weinstein to three months of incarceration at Wormwood Scrubs prison. Following this, Weinstein was handed over to military authorities, who promptly court-martialled him at Bedford Barracks in November 1916 for disobeying conscription orders. In an ultimately pointless exercise, the military sentenced Simy to six months of hard labour before releasing and re-arresting him for resisting conscription. This cycle continued until Weinstein was permanently released in 1919—well after the end of fighting.[32]

Imprisonment and the social fallout of being a conscientious objector placed a severe strain on the Whitechapel Boys and their families and friends. Serving time in Wandsworth, Bedford, and Reading, Simy Weinstein later published a volume of poetry titled *Chains* (1920) on this experience. It is possible that he was also involved in hunger strikes during his imprisonment. Fellow Whitechapel Boy John Rodker first escaped arrest and hid with the poet R.C. Trevelyan but was soon caught and imprisoned in Dartmoor. Similar to Weinstein, Rodker later used his experiences to anonymously write *Memoirs of Other Fronts* (1932). By contrast, Clara's brother Jonas evaded imprisonment at Cambridge.[33]

Life was not easier outside prison. Following the outbreak of war, Clara had initially volunteered to teach art at a school for boys but a pregnancy and Simy's incarceration in 1916 prompted her to move to the countryside so that she could be closer to her husband. Living with the wife of another conscientious objector, she caused a stir among locals with her short hair and habit of walking barefoot. The couple's first child, Theodora, was born at a wealthy friend's home in Warwickshire while Stephen was imprisoned in 1917. Theodora's birth increased Clara's commitment to vegetarianism and other reform movement ideals such as eating raw and 'natural' foods and nudism. It also prompted Clara to move back to London where a wealthy sister of Samuel—probably Mary—provided her with accommodation in Highgate. In between prison visits, Clara designed toys and supported pacifist, vegetarian, and suffragette causes.[34]

The end of the war brought not only an end of imprisonment but also a severance of artistic and ideological ties between the former Whitechapel

[32] Peace Pledge Union "Remembering" project; and Winsten Autobiography, 68–70.

[33] Macdougall, "Something Is Happening There," 140, Macdougall, "Whitechapel Girl," 112.

[34] Winsten Autobiography, 72–85.

Boys. Following Simy's release, Samuel and Clara anglicised their names to Stephen and Clare Winsten and briefly moved to Bournemouth before finding it necessary to move back to London where they could rely on friends and family for support.[35] After moving in with Russian communist contacts of Stephen in Bedford Park, the family soon relocated to a studio close to Kensington Parks. Sporadic patronage and picture sales could not alleviate long-term money worries, which were exacerbated by the fact that Stephen found it hard to gain employment as a teacher due to his status as a conscientious objector.[36]

Despite their financial woes, the Winstens' ties to London's artistic and intellectual milieu remained excellent. In Kensington, they enjoyed regular contact with Lucien Pissarro and his family. Despite mixed reviews, Stephen contributed to journals like *Voices* and the Yiddish *Renesans* and organised cultural events at Toynbee Hall. Meanwhile, Clare's work was displayed at various salons and exhibitions.[37] In 1920, the Winstens—with Clare already pregnant—temporarily moved to a relative's Gothic inspired summer house before returning to London in preparation for the birth of Ruth. Ruth herself was born prematurely at seven months in Fulham in London. According to Clare, the birth was dramatic, with Ruth barely surviving the first hours of her life. Ruth's first visitors were Lucien Pissarro and his wife. Clare's drawings of the first seven days after Ruth's birth have unfortunately disappeared (Image 2.2).[38]

Things quickly became crowded at the family's Kensington flat. Clare and Stephen slept in the upstairs studio while the two children shared a wide bed with an au pair. Ahead of the birth of their third child, Christopher Blake Winsten in 1923, the family decided to leave the metropolis for Ebenezer Howard's second Garden City in Welwyn, where they purchased a small house with money from Clare's brother and a wealthy friend. In Welwyn, the Winstens enjoyed close contact with the many artists, civil servants, and intellectuals who had moved there; helped open a local Montessori school; and organised lectures by well-known friends, including George Bernard Shaw.[39]

[35] Winsten Autobiography, 102–103.

[36] Macdougall, "Whitechapel Girl," 112–113; Winsten Autobiography, 105–106; 128; his right to vote was also curtailed by the 1918 Representation of Peoples Act.

[37] Winsten Autobiography, 92, 97, 99–, 104–109, 114–116, 128; MacDougall, "Whitechapel Girl," 113.

[38] Winsten Autobiography, 110, 112, 115–116, 138–139.

[39] Winsten Autobiography, 116, 138–139, 140, 145–146, 154, 117, 119–121, 123–133, 136D, 161–163.

Image 2.2 Winsten Family Portrait of Clare, Theodora, Ruth, and Stephen, Photograph ca. 1922 (image courtesy of Jonathan Harrison)

For the sake of their children's education, Clare and Stephen decided to move back to Hampstead in North London towards the end of the 1920s. All three children were placed in local schools, and Christopher managed to win a stipend for University College School, from which he went on to study mathematics at Cambridge and become a leading probability theorist at the University of Essex. Clare and Stephen continued to cultivate contacts to London's artistic circles and co-edited the magazine *To-morrow* with Hugh Walpole and Bertrand Allison.[40] However, the couple's own artistic output became increasingly sporadic. As a mother of

[40] Winsten Autobiography, 135–136; "Winsten, Clare & Stephen," in W. Rubenstein and Michael Jolles (eds.), *Palgrave Dictionary of Anglo-Jewish History*.

three, Clare complained that living among Britain's artistic elite was not the same as being recognised as an artist: "It seemed as though success was not for me. Always on the brink but never there. (...). Had I been mistaken in myself? Why had I been given a scholarship, why had I been praised by many for my work? It is a humiliating experience, that of being cast out of the group of artists of my generation."[41]

A major exception to this perceived lack of recognition was Clare's commission to portray Mohandas (Mahatma) Gandhi on his 1931 visit to Britain. Clare had been selected as portraitist on the grounds of the Winstens' pacifist vegetarian credentials and ties to local Indian activists. Although she denied being a portrait artist, Clare also portrayed other famous individuals and family acquaintances including Ebenezer Howard, Ezra Pound, Benjamin Britten, W.H. Auden, Dmitri Shostakovich, Maria Montessori, and family friend George Bernard Shaw. The portraits not only proved to be some of her best-known works but also provided much-needed income for the Winsten household.[42]

Shaw in particular would become an important patron of and influence on Clare, Stephen, and their children. Ahead of the outbreak of the Second World War, the Winstens briefly moved to Wales and Huntingdonshire before settling in Ayot St Lawrence in Hertfordshire in 1940, where they became Shaw's neighbours.[43] The family's intimate contact with Shaw created economic opportunities and intensified their engagement with the synthesist animal and human ethics of Edwardian reform movements. As a playwright and public figure, Shaw personified the integration of socialist, pacifist, vegetarian, and humanitarian values. After coming to London in 1876, the Irish-born dramatist had become immersed in radical politics, converted to vegetarianism around 1881, and joined the newly formed socialist Fabian Society in 1884. Eleven years later, he co-founded the London School of Economics—where Samuel Weinstein briefly took courses—with fellow Fabians Sidney and Beatrice Webb and Graham Wallas.[44]

Shaw's vision of society's moral evolution via vegetarianism and peaceful socialism was influenced by his friend Henry Stephens Salt. A former assistant headmaster of Eton, Salt campaigned for the humane treatment of all creatures and co-founded the Humanitarian League in 1891. Drawing on

[41] Winsten Autobiography, 149–150.

[42] Winsten Autobiography, 146, 154; Ayad, "The Winstens of Whitechapel"; Dickson and MacDougall, "The Whitechapel Boys," 34; notable sculptures include "Joan of Arc" in Ayot and "Mother and Child" (1968) at Toynbee Hall.

[43] Winsten Autobiography, 154; Ayad, "The Winstens of Whitechapel".

[44] Michael Holroyd, *Bernard Shaw* (London: Random House, 2011), 51–53, 102–107, 126–127, 291

support from a wide range of influential friends, including Arts and Crafts socialist William Morris and anarcho-communist Prince Kropotkin, he campaigned for criminal reform, education, sanitation, and decolonisation alongside vegetarianism, ending blood sports, and preventing the use of animals for fashion. In 1892, he published *Animals' Rights*, which drew on a mix of evolutionary argumentation, natural history, materialism, and Benthamite philosophy to argue that all animals should be protected from unnecessary suffering. According to Salt, there was no dichotomy between 'nature' and 'society' and between the struggle to improve human conditions and for a more humane treatment of animals.[45]

The Winstens' vegetarian values and pacifist credentials made them perfect neighbours for Shaw. Spending much of the 1940s in the orbit of the ageing playwright, the family also enjoyed intimate contact with the many likeminded Edwardian reformers visiting Ayot. Stephen in particular profited from this proximity. Commuting to London to fire watch during the Blitz,[46] he later published a biographical account of his wartime *Days with Bernard Shaw*.[47] In 1946, he also edited a *Festschrift* for Shaw including contributions from political, scientific, and literary luminaries such as J.B. Priestley, H.G. Wells, John Maynard Keynes, and Aldous Huxley.[48] Drawing on Shaw's archived personal correspondence with Henry Salt, who had died in 1939, Stephen also published *Salt and His Circle* in 1951.[49] The book adopted a light-hearted tone to revisit—through the eyes of Shaw—crucial episodes of Salt's life and humanitarian struggles via invented dialogues, letter excerpts, and psychological characterisations of key figures. Although it had no lasting impact on scholarship, the often-idolising tone of *Salt and His Circle* is indicative of the extent to which Stephen shared Salt's fusion of socialist, pacifist, vegetarian, and evolutionary concepts of interspecies kinship (Image 2.3).[50]

[45] Brett Clark and John Bellamy Foster. "Henry S. Salt, socialist animal rights activist: An introduction to Salt's A Lover of Animals," *Organization & Environment* 13/4 (2000), 468–473; Simon Wild, "Henry S. Salt," *Henry S. Salt Society*, https://www.henrysalt.co.uk/life/biography/ [01.05.2020].

[46] Stephen Winsten, *Days with Bernard Shaw* (London: Readers Union/ Hutchinson, 1951), 118.

[47] Winsten, *Days with Bernard Shaw*.

[48] Rod Preece, *Animal Sensibility and Inclusive Justice in the Age of Bernard Shaw* (Vancouver: UBC Press, 2011), 19.

[49] Stephen Winsten, *Salt and His Circle* (London: Hutchinson & Co. Ltd, 1951).

[50] See Stephen Winsten's discussion of vegetarianism, evolutionary theory, animal rights and kinship, and the humanitarian league, *Salt and His Circle*, 87–88, 102, 116–118, 127, 131–134, 182.

Image 2.3 George Bernard Shaw by Clare Winsten, pencil on paper laid on board, 1949, NPG 6891 (image courtesy of National Portrait Gallery)

Personal and intellectual ties between the neighbours remained good for most of the decade. According to biographer Anthony Gibbs, Shaw considered the Winstens "a talented Bohemian family who offered him what no one else in the village could, intelligent conversation."[51] Shaw engaged in active patronage of the entire family. In 1947, he commissioned Clare to create a statue of *Joan of Arc* for Shaw's garden around which his ashes were later scattered and asked her to illustrate his *Buoyant Billions: A Comedy of No Manners in Prose* in 1949.[52] The Winsten children

[51] Quoted according to Anthony Matthews Gibbs, *A Bernard Shaw Chronology* (London: Palgrave, 2001), 393.

[52] Gibbs, *A Bernard Shaw Chronology*, 392–393; Shaw allegedly refused to pay for Clare Winsten's painted portrait of him; there was also a short period of estrangement following disagreement between Shaw and Clare Winsten about the placement of the statue in the garden, J.P. Wearing, *Bernard Shaw and Nancy Astor. Selected Correspondence of Bernard Shaw* (Toronto et al.: University of Toronto Press, 2005), 202, 207; see also Winsten Autobiography, 222-225; 263–265.

also received support: Shaw contributed £2000 to Christopher's Cambridge education, trained Ruth in drama, and helped Theodora, a Prizeman in Stage Design at the Slade, secure her first commission as designer for his *Buoyant Billions* play.[53] He also dedicated a pamphlet to Theodora, which was posthumously edited and illustrated by Stephen and Clare (*My dear Dorothea: a practical system of moral education for females, embodied in a letter to a young person of that sex*).[54]

It was only in 1949—one year ahead of Shaw's death—that neighbourly relations soured. Having previously ignored accusations that the Winstens were making a living out of him, significant inaccuracies in Stephen's writings forced Shaw to publish a disclaimer about *Days with Bernard Shaw* in the *Times Literary Supplement*.[55] Shaw also refused to pay for an extension of the Winstens' lease at Ayot. Although he allegedly later regretted it, the decision forced the cash-strapped family to move to Oxford.[56] For Stephen and Clare, the move marked the end of easy access to Britain's cultural elite. Forty years after joining the Whitechapel Boys, the couple's 'downstart' careers had peaked. While the Winstens never achieved the artistic fame of friends like Rosenberg, their lived pacifist and vegetarian values and social milieu left an indelible mark on their children.

[53] Dan H. Laurence, *Bernard Shaw Theatrics. Selected Correspondence of Bernard Shaw.* (Toronto et al.: University of Toronto Press, 1995), 231; Michael Holroyd, *Bernard Shaw. 1918–1950. The Lure of Fantasy* (London: Chatto & Windus, 1991), 467.

[54] Macdougall, "Whitechapel Girl," 113–14; Gibbs, *A Bernard Shaw Chronology*, 392–393.

[55] George Bernard Shaw, "Conversation Pieces," *TLS* (15.01.1949), 41.

[56] Holroyd, *Shaw. Lure of Fantasy*, 470.

CHAPTER 3

Becoming an Activist: Ruth Harrison's Turn to Animal Welfare

Stephen and Clare's 1949 move from Ayot to Oxford coincided with their daughter Ruth's decision to abandon a potential artistic career in favour of paid employment. Her turn away from the arts in search of economic stability did not dampen Ruth's commitment to the reform values of her family or her already strong commitment to bearing witness to these values as an activist.

According to her mother's autobiography, Ruth's childhood had been a happy one. Living in immediate proximity to a Friends' (Quaker) Meeting House in Welwyn, Ruth was raised according to the 'nature-oriented' tenets of the contemporary reform movement:

> Our babes could be in the garden all day long, bare footed and most often bare. We had created a huge sand centre shielded by loganberry bushes and with pails and spade they were perpetually busy. On the lawn we built brick steps sapped with a stretch of grass so that they could climb and roll down on the other side. Soon we found the ideal roundabout cum see-saw and fixed it securely on the lawn.[1]

[1] Winsten Autobiography, 146; see also: 119 & 153.

C. Kirchhelle, *Bearing Witness*, Palgrave Studies in the History of Social Movements,
https://doi.org/10.1007/978-3-030-62792-8_3

Her parents' educational principles proved contentious among neighbours, who complained to the local police about the children's nudity and vegetarian diet, which was allegedly starving them.[2] This did little to change Clare and Stephen's convictions. Ruth was subsequently enrolled in Welwyn's Montessori Community School, which her parents had co-founded, and attended the many cultural events organised by the Winstens—including lectures by Shaw.[3]

When the family moved back to Hampstead in London in the late 1920s, Ruth was placed in Parliament Hill School on the other side of Hampstead Heath, where she developed a strong interest in drama, arts—and animals.[4] According to her mother, this latter interest had been pronounced from early on and resulted in the repeated adoption of new animals by the household. During a vacation in Devon, Ruth:

> helped the farm lady in the feeding of chickens, grooming the ponies, and otherwise helping the owner of this farm. At the end of our month's stay, we were seen off by the lady of the farm who handed a small cake box to Ruth. Thinking this a packet of sandwiches for our journey we thanked her and left. While in the train we heard a 'cheep cheep' coming from the box. To our amazement there were no sandwiches but a chicken![5]

Ruth explained that the chicken had been neglected and kept in the barn for warmth where she had fed it all day: "This was an embarrassment because we lived in London. However to please little Ruth we took it home and it lived in the studio, and to keep it company Ruth persuaded us to buy a ginger kitten."[6]

After finishing secondary school, Ruth pursued her interest in drama and enrolled as an English major in Bedford College, London University, in summer 1939. Because of the outbreak of war, her college was evacuated to Cambridge.[7] Occurring parallel to her parents' move to Ayot, this move was highly significant in terms of Ruth's decision to join

[2] Winsten Autobiography, 132.

[3] Winsten Autobiography 119–120, 117, 161–162, 133.

[4] Ryder, "Harrison, Ruth (1920–2000)"; her mother recounts Ruth's shock at seeing a cat killing a rabbit in Sussex, Winsten Autobiography, 134A.

[5] Winsten Autobiography, 139–140/ 166–167.

[6] Winsten Autobiography, 139-140/ 166–167.

[7] Ryder, "Harrison, Ruth (1920–2000)".

the local theist Society of Friends (Quakers).[8] As the daughter of a vegetarian New Woman and a conscientious objector, many core tenets of British Quakerism's "middle-class radicalism"[9] would have been familiar to her. Founded in the northwest of England in 1652, Quakerism holds that there is something of God in everyone. Because of this belief, Quakers place a strong emphasis on social justice, equality, and pacifism. Another important part of Quakerism consists of living faith through peaceful action and registering disapproval against grievances.[10] Also known as bearing witness, this practice of peaceful protest made Quakers play highly visible roles in nineteenth- and twentieth-century anti-slavery, temperance, and social justice movements. Some Quakers also extended the principle of the sanctity of life and non-violence to animals and supported the antivivisection movement. Founded in 1902, the Friends' Vegetarian Society in particular dedicated itself to bearing witness against cruelty to animals and spreading vegetarianism and other reform principles among fellow Quakers.[11] It is not difficult to see why this fusion of compassion for humans and animals, non-violent mode of bearing witness to injustice, and tradition of middle-class radicalism made Quakerism appeal to Ruth.

The fact that her conversion occurred in Cambridge was doubly significant because it brought Ruth into the vicinity of a leading British expert in animal behaviour. Prior to becoming a fully 'convinced' (converted) Friend, Ruth would have regularly attended meetings of Cambridge's local Society of Friends.[12] While doing so, she likely met William Homan Thorpe. Thorpe was an influential Cambridge entomologist and animal behaviourist. During the early 1940s, he was not only committing to Quakerism himself but also developing a synthesist fusion of Darwinian evolution and Christianity centring on animal cognition (see Chap. 4). This research would provide important scientific support for Ruth's

[8] Kendall, "Ruth and the Ruthless," 2; Van De Weerd and Sandilands, "Bringing the Issue of Animal Welfare to the Public," 404.

[9] Frank Parkin, *Middle Class Radicalism: The Social Bases of the British Campaign for Nuclear Disarmament* (Manchester: Manchester University Press, 1968), 2–5.

[10] Pink Dandelion, *The Quakers: A Very Short Introduction* (Oxford: Oxford University Press, 2008), 1–2, 36.

[11] Dandelion, *The Quakers*, 33–35; Julia Twigg, *The Vegetarian Movement in England, 1847–1981: A Study In The Structure Of Its Ideology* (London: Dissertation London School of Economics, 1981), Chapter 7.IV.

[12] "Chapter 11: Membership", *Quaker Faith and Practice* 5th edition, https://qfp.quaker.org.uk/chapter/11/ [08.11.2019].

campaigning in the 1960s. However, in the short term, Thorpe likely also facilitated Ruth's first activist experience in his role as chairman of Cambridge's Pacifists' Service Bureau, which helped local Quakers and other conscientious objectors find wartime work in organisations like the Friends' Ambulance Unit (FAU).[13]

Founded in accordance with the Quaker peace testimony, the FAU had provided medical relief for combatants and civilians of all nationalities during the First World War. In September 1939, the FAU was refounded by former members. Early FAU work was far from glamorous. Despite being eager to help, most FAU volunteers had limited medical experience and found themselves changing linen, cleaning wards, and performing basic nursing duties. Starting in September 1940, the Blitz changed this situation. Headquartered in London's Whitechapel neighbourhood, the FAU provided important emergency aid and rapidly expanded its activities, first throughout Britain and then abroad. In total, the FAU trained 1300 volunteers as ambulance drivers, medical orderlies, and unqualified nurses. Women were allowed to aid FAU relief work from October 1940 onwards. By 1941, they were permitted to train as full FAU members. Overall, 97 women would train with and work for the FAU.[14]

Ruth was one of these 97 women. Aged 23, she joined the FAU in 1943 and attended a 14-day FAU training course before starting work as an unqualified nurse in Hackney Hospital.[15] According to a later interview, her first day of work left a marked impression on her:

> after seeing festering ulcers, [I] went home and sobbed all night. 'At dawn I said, 'Come on, girl, get some guts,' so I went back and made myself work all day on the worst ulcers, the nastiest whitlows. It was a slog, but searing.[16]

[13] R.A. Hinde, "William Homan Thorpe. 1 April 1902—7 April 1986," *Biographical Memoirs of Fellows of the Royal Society* 33 (1987), 625; Oral History Interview Donald Broom (04.07.2014).

[14] A. Tegla Davies, *Friends Ambulance Unit. The Story of the F.A.U. in the Second World War 1939–1945* (London: George Allen and Unwin Limited, 1947); "Quakers in Action. Women in the FAU", *Quakers in the World*, http://www.quakersintheworld.org/quakers-in-action/329 [08.06.2016].

[15] Library of the Society of Friends [in the following LSF], Friends' Ambulance Unit [in the following FAU] Record Cards, Ruth Winsten, 1.

[16] "Debt of honour", *Observer*, 02.09.1973, 40; while the *Observer* claims that Harrison was unqualified, Harrison's son, Jonathan, notes that she passed her nursing exam with a score of 99 per cent (Correspondence with Jonathan Harrison 29.08.2015); Richard Overy, "Pacifism and the Blitz, 1940–1941," *Past & Present* 219 (2013), 217–18.

Although the most intensive phase of the Blitz had already passed, work within metropolitan hospitals was not without its dangers. British cities were still subjected to occasional bombing, and nurses like Ruth could fall victim to hospital-acquired infections. After four months of work at Hackney Hospital and in Lichfield, Ruth contracted scarlet fever and was placed in Burton Isolation Hospital before being sent to convalesce with her parents in Ayot. She returned to work within two months and spent the period between 1944 and mid-1945 working at St Mary's Islington Hospital, Hackney Hospital, Queen Elizabeth Hospital, and in FAU administrative hubs in London (Images 3.1 and 3.2).[17]

With pressure on British hospitals declining after D-day in 1944, many FAU volunteers began to look for alternative postings. Postings abroad

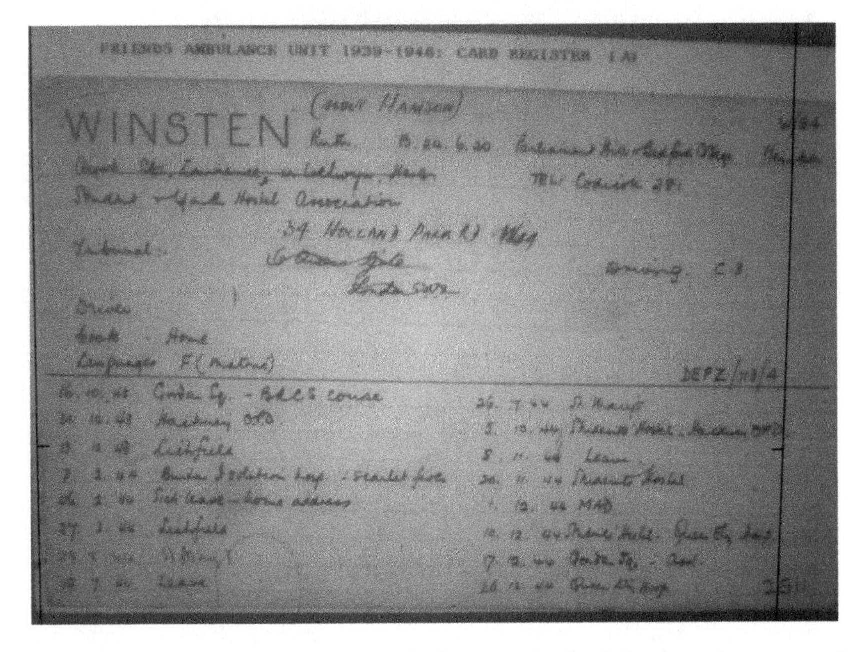
FRIENDS AMBULANCE UNIT 1939–1946: CARD REGISTER (A)

WINSTEN

Image 3.1 Picture in Friends' Ambulance Unit Card Register (courtesy of Friends' Library, London)

[17] LSF, FAU Record Cards, Ruth Winsten, 1.

Image 3.2 Ruth Winsten Personnel File (courtesy of Friends' Library, London)

were particularly prestigious. Since the early 1940s, FAU ambulance sections had been active in Finland, Norway, Greece, China, Syria, India, and then in Italy, France, and Germany. Many of these sections were mixed gender. However, due to their integration with military units, FAU sections in Northern Europe were initially all-male. This was difficult to stomach for the daughter of a New Woman. Writing to the FAU's Executive Committee in January 1945, Ruth drew attention to female FAU members' desire to aid European relief efforts. After a meeting with female members, the executive committee agreed to change course. By June 1945, the first female FAU workers joined male colleagues in continental Displaced Persons camps.[18]

[18] LSF, FAU Executive Committee Minutes Feb. 1945 to Dec. 1945, Minutes of Meeting of the Executive Council, 01.02.1945, item 3153, 1; "Quakers in Action. Women in the FAU", *Quakers in the World*; "Quakers in Action. FAU in WWII: Civilian Relief Work in

As the daughter of ethnic Jews, Ruth made the remarkable decision to aid FAU relief efforts in occupied Germany.[19] Having attended a German language course in June 1945, she was supposed to join one of the first mixed FAU units in early July. However, sickness delayed her departure.[20] By the time of her arrival in August 1945, there were 150 FAU members working in Germany.[21] As one of only 57 female FAU members to work abroad, Ruth was assigned to FAU Section 133. Between 1945 and 1946, she was deployed in Bochum in the Ruhr area and in the small town of Husum close to Schleswig-Holstein's Danish border, which had been the site of a concentration camp outpost for foreign labourers.[22]

Regular reports from FAU Sections give insight into Ruth's work. Initially, FAU Section 133 provided relief and repatriation services for former concentration and labour camp inmates and non-Germans. Following the relaxation of fraternisation restrictions in mid 1945, the FAU also helped to establish refugee camps and provide education, food, and medical relief for German civilians.[23] Conditions were difficult. In their reports, FAU members expressed concern about the humanitarian situation in bombed-out cities. With only limited access to food, sanitation, and medical services, diseases spread quickly and mortality was high. Liaising with local volunteers, FAU workers were also keenly aware that many of their German counterparts had until recently formed part of the Nazi machinery. However, in accordance with its Quaker convictions, the FAU hoped that reconstruction and relief would create a new peaceful Germany. As a consequence, it played an important role in raising awareness about the appalling living conditions of German children and other vulnerable people amongst local Allied military authorities and in Britain. Setting out its education programme for Duisburg youths and unaccompanied children living in bunkers in December 1945, FAU Section 5 warned that rhetorical commitments to a new Germany were not enough in a situation "where living is so often a matter of cold, (…) and hungry bellies."[24] Ruth's FAU section

Mainland Europe", *Quakers in the World* (http://www.quakersintheworld.org/quakers-in-action/295 [08.06.2016]).

[19] Ryder, "Harrison, Ruth (1920–2000)"; Van De Weerd and Sandilands, "Bringing the Issue of Animal Welfare to the Public," 404.

[20] LSF, FAU Record Cards, Ruth Winsten, 2.

[21] "Quakers in Action. Women in the FAU," *Quakers in the World.*

[22] "Quakers in Action"; Ryder, "Harrison, Ruth (1920–2000)"; Davies, *Friends Ambulance Unit*, 435.

[23] "Quakers in Action: FAU in WWII: Civilian Relief Work in Mainland Europe," *Quakers in the World.*

[24] LSF, FAU Records, Folder "German Relief", "Young People from 0–18 Years in Duisburg, Germany", 2.

pressured British authorities to provide fuel for hospitals and nurseries, monitored the situation of displaced children and German refugees in Denmark, and reported on dramatic hikes of infant mortality.[25] The FAU also assessed the ideological state of German youth organisations, re-established contact with German Friends, and monitored evangelical churches.[26]

Despite its success as a relief organisation, the FAU had not been designed for prolonged post-war activity. About a year after Nazi Germany's defeat, it began to wind down its international work. While a small number of personnel continued to provide relief on the continent, the majority of FAU members were withdrawn and resigned from active duty by summer 1946.[27] After nearly three years of FAU work and ten months in Germany, Ruth returned to Britain in mid-June 1946 and resigned from the FAU in mid-July.[28] The war undoubtedly marked a turning point in her life. In addition to converting to Quakerism, her decision to join the FAU and volunteer in Germany—despite her Jewish family background—highlights how far she would go to bear witness to humanitarian causes in which she believed.

Back in Britain, the now 26-year-old Ruth continued her university studies and enrolled at the prestigious Royal Academy of Dramatic Art (RADA). Her sister, Theodora, also a RADA graduate, studied stage design down the road at their mother's old school, the Slade. Ruth showed talent. Before she joined RADA, George Bernard Shaw coached her to throw her voice in the church at Ayot St Lawrence.[29] At RADA, Ruth conducted her own production of J.B. Priestley's *An Inspector Calls*.[30] Priestley himself complimented Ruth on the production. In the same year, Harrison also won RADA's one-act-play competition.[31] Recognising her promise, Shaw unsuccessfully recommended Ruth as producer for the 1949 Malvern Festival production of his *In Good King Charles's Golden Days* (Image 3.3):

[25] LSF, FAU Records, Folder "German Relief", FAU Relief Section 133, Harold Cadows, "Report on fuel supply in Bochum" (19.02.1946); Beatrice Thrift, "Refugee Camps in Denmark" (01.03.1946); Pip Turner, "Population Statistics for Bochum for the year of 1945"; Idem., "Statistics of Infectious Diseases in Bochum January-February 1946" (28.03.1946); Idem, "Population Statistics for Bochum for the month of March 1946" (04.05.1946); "Report on the administration of education in Gelsenkirchen" (29.12.1945).

[26] LSF, FAU Records, Folder "German Relief", The Evangelical Church in Bochum (report); Observations on German Political Movements (report).

[27] Davies, *Friends Ambulance Unit*, 439–440.

[28] LSF, FAU Record Cards, Ruth Winsten, 2.

[29] Correspondence with Jonathan Harrison (29.08.2015)

[30] Van De Weerd and Sandilands, "Bringing the Issue of Animal Welfare to the Public," 404.

[31] Ryder, "Harrison, Ruth (1920–2000)"; Correspondence with Jonathan Harrison (21.01.2015); Dan. H. Laurence, *Theatrics*, 231.

Image 3.3 Theodora and Stephen Winsten, George Bernard Shaw, Ruth Winsten, Devdas Gandhi (son of Mohandas Gandhi) and Clare Winsten at Ayot St. Lawrence (ca. 1949) (image courtesy of Jonathan Harrison)

> She is in every way a desirable person in the theatre, and understands that my plays are essentially religious and serious, however entertaining they may be, and no matter how many laughs they may get when the actors don't play for them. I could trust her to produce much more hopefully than these 'Where's your murder?' chaps.[32]

Despite these glowing endorsements, Ruth decided not to pursue a career in the dramatic arts.[33] Coinciding with her parents' departure from Ayot, she joined the architectural firm Harrison and Seel. In 1954, she married the company's senior partner Dexter 'Dex' Harrison (1909–1987).[34] Dex had studied at Leeds School of Architecture and moved to London in the 1930s. During the war, he had worked for the Ministry of Works and authored a major *Survey of Prefabrication* (1945) in preparation for British post-war reconstruction. After 1945, he was

[32] Quoted according to Laurence, *Theatrics*, 230.

[33] Ena Kendall, "Ruth and the Ruthless," 2.

[34] "Dex Harrison—Basic Biographical Details", *Dictionary of Scottish Architects, Architect Biography Report* http://www.scottisharchitects.org.uk/architect_full.php?id=206027 [20.12.2014].

chief architect of the 1951 Festival of Britain's pleasure gardens and designed the new theatre and other buildings around Battersea Park.[35] Although Dex remained a meat-eater, his 1954 marriage to Ruth was happy. In 1955 and 1956, the couple had two children, while Ruth complied with contemporary *mores* and became a housewife in a "calm studio house in [London Kensington] with ... two pianos but only one cat"[36] (Image 3.4).

Image 3.4 Dex Harrison's pleasure gardens at the 1951 Festival of Britain (image courtesy of Jonathan Harrison)

[35] "Dex Harrison—Basic Biographical Details"; "Obituary—Dex Harrison," *The Times* (15.01.1988), 14.

[36] "Debt of honour", *Observer*, 02.09.1973, 40; for studies on the effect of marriage and motherhood on women's careers during the 1950s and 1960s, see Sarah Aiston, "A Good Job for a Girl? The Career Biographies of Women Graduates of the University of Liverpool Post-1945," *Twentieth Century British History*, 15/4 (2004), Carol Dyhouse, "Family Patterns of Social Mobility through Higher Education in England in the 1930s," *Journal of Social History*, 34/4 (2001), Dolly Smith Wilson, "A New Look at the Affluent Worker: The Good Working Mother in Post-War Britain," *Twentieth Century British History*, 17/2 (2006).

Life did not stay quiet for long. Like many Quakers, Harrison supported the Campaign for Nuclear Disarmament (CND). Officially founded in February 1958, the CND united many different groups with its call for unilateral nuclear disarmament. Its foundation was triggered by concerns about nuclear testing and a new generation of hydrogen bombs, whose destructive power far exceeded previous nuclear weapons. Supported by a broad alliance of Labour activists, church leaders, and the Peace Pledge Union, the CND became the largest British extra parliamentary organisation after 1945. In 1961, its annual Aldermaston march attracted ca. 150,000 people. The movement also germinated new forms of protest. Founded in 1960, the Committee of 100 was led by the charismatic philosopher Bertrand Russell, an acquaintance of Harrison's maternal uncle Jonas Birnberg, and coordinated civil disobedience in the form of nonviolent direct action like sit-ins.[37]

Although CND demands for unilateral disarmament were ultimately ignored, the peace movement provided important impulses for post-war environmentalism. In Britain, many CND-campaigners soon expanded their protest to encompass destructive non-nuclear technologies and the global threat posed by environmental degradation. In doing so, these campaigners contributed to what some historians describe as the 1960s' shift from an anthropocentric to an ecocentric view of the world. Early British environmentalist and CND activists likewise often shared a yearning for an allegedly simpler past and a distrust of post-war technologies.[38]

Quakers played a prominent role in both the transatlantic peace and environmentalist movements. In a 1965 survey of 368 British CND members, a remarkable 10 per cent of respondents described themselves as Quaker (14 per cent were Church of England, 15 per cent atheist, 15 per cent humanist, and 19 per cent agnostic)[39] even though there were only 15,000–20,000 registered Quakers in Britain.[40] Focusing on the US, historian Frank Zelko has highlighted the importance of Quaker convictions for early Greenpeace activists like Irving and Dorothy Stowe. In an interesting

[37] Oral History Interview Donald Broom (04.07.2014); on Christians and the CND see Adam Lent, *British Social Movements since 1945*, 41–45, Veldman, *Fantasy*, 115–200; Burkett, "The Campaign for Nuclear Disarmament," 627–632; Harrison's parents might have participated in the anti-war rallies on Trafalgar Square in 1914; Macdougall, "Whitechapel Girl," 112.

[38] Veldman, *Fantasy*, 115–180; Burkett, "The Campaign for Nuclear Disarmament," 626–627.

[39] Parkin, *Middle Class Radicalism*, 27.

[40] James Chadkirka, *Patterns of Membership and Participation among British Quakers, 1823–2012* (MA thesis, University of Birmingham, 2014), 58.

parallel to Ruth Harrison, the Stowes grew up as liberal Jews on the US East Coast during the interwar period but were drawn to Quaker humanism during the late 1940s. Following the US' use of nuclear weapons against Japan, the Stowes participated in anti-nuclear peace movement campaigns and went on to form the core of Greenpeace in the 1960s. According to Zelko, the concept of bearing witness—that is, "registering one's disapproval of an activity and putting moral pressure on the perpetrators"[41]—made the Stowes and other Quakers inherently attracted to post-war Civil Rights, anti-nuclear, and environmental movements.[42] To this day, environmentalist values and eco-spiritualism form an important part of global Quaker thinking about non-ostentatious plainness, sustainability, and social justice.[43]

In the British context, Quakers' commitment to environmentalist values and peacefully bearing witness to humanitarian values is key to understanding Ruth Harrison's decision to write *Animal Machines*. When the Crusade Against All Cruelty to Animals pushed a leaflet against 'factory farming' through her door around 1960,[44] they spurred to action a politically compatible and well-connected Quaker, who knew the power of civic activism. The leaflet itself contained a series of disturbing images and descriptions of animals' living conditions under intensive housing and slaughtering practices in abattoirs. Calling for more legislative protection, it exhorted readers to join the Crusade and write to their MPs and newspapers in protest.[45] According to a 1990 interview, Ruth Harrison's initial reaction was to do nothing. As a "life-long vegetarian," she felt that a campaign for farm animals did not concern her: "But 'in doing nothing I was allowing it to happen'"[46] (Image 3.5).

[41] Zelko, *Make It a Green Peace!*, 13.

[42] Zelko, *Make It a Green Peace!*, 11–15; 20–27; 32.

[43] Peter Jeffrey Collins, "The development of ecospirituality among British Quakers," Ecozon@ 2/2 (2011); Timothy Burdick and Pink Dandelion, "Global Quakerism 1920–2015," in Stephen W. Angell and Pink Dandelion (eds.), *The Cambridge Companion to Quakerism* (Cambridge: Cambridge University Press, 2018), 49–66; Emma Jones Lapsansky, "The Changing World of Quaker Material Culture," in Stephen W. Angell and Pink Dandelion (eds.), *The Cambridge Companion to Quakerism* (Cambridge: Cambridge University Press, 2018), 147–158.

[44] There are conflicting dates regarding the leaflet; whereas Helen van de Weerd and Victoria Sandilands claim that Harrison received the leaflet in 1961, Ena Kendall holds that Harrison received in the late 1950s; in 1990, the *Guardian* reported that Harrison received the leaflet in 1960; Van De Weerd and Sandilands, "Bringing the Issue of Animal Welfare to the Public," 405; Kendall, "Ruth and the Ruthless," 2; Colin Spencer and Spike Gerrel, "A rare breed at the factory farm", *Guardian*, 03.11.1990, A19.

[45] Van De Weerd and Sandilands, "Bringing the Issue of Animal Welfare to the Public," 407.

[46] Quoted according to Colin Spencer and Spike Gerrel, "A rare breed at the factory farm", *Guardian*, 03.11.1990, A19.

WHAT YOU CAN DO

- Refuse to buy forced white veal and broiler chickens and tell the shopkeeper why. Cut-price chickens can only be obtained by broiler methods.
- Do all you can to avoid buying battery eggs. Ask for FREE RANGE eggs or buy DANISH. Tell the shopkeeper you prefer British eggs but will not buy while they are produced by battery methods.
- Write to your Member of Parliament, House of Commons, London, S.W.1. protesting against the broiler and battery systems and ask him to take action on the matter. There are attempts to give the Minister of Agriculture, Fisheries and Food and the Secretary of State for Scotland power to make regulations concerning intensive methods of food production, i.e. the broiler calf and chicken industries, but this would allow the system to continue even if in modified form.
- We want an amendment to the PROTECTION OF ANIMALS ACT, 1911, to make these systems **illegal**. We appeal to you in the name of sanity to write to your M.P. asking him to support us in this.
- Ask your M.P. to agitate at once for all "broiler" chickens, forced white veal and battery eggs to be marked as such so that you, the public, can make the choice you are entitled as free individuals to make when buying your food.
- Write to the national and local papers about it and keep writing. Talk about it in your local societies and church organisations and when you go shopping.
- Join our national campaign against these evils as announced in THE DAILY MIRROR of December 8, 1960.

Remember **.. "All that is necessary for the triumph of evil is that good men do nothing"** —*Burke*.

Issued by

CRUSADE AGAINST ALL CRUELTY TO ANIMALS

3, Woodfield Way, Bounds Green Rd., London, N.11.

in co-operation with

CAPTIVE ANIMALS' PROTECTION SOCIETY

Further copies of this leaflet can be obtained from the above address free of charge but DONATIONS to the campaign will be gratefully received. Cheques and postal orders should be made payable to HUMANE FARMING CAMPAIGN and crossed "& Co."

Cheap food? YES!

BUT IS IT GOOD FOOD?

"Farmer & Stockbreeder" photograph.

"BROILER" CALVES – in prison for life!

The Dutch method of rearing calves for veal has recently been introduced into this country and is being developed despite public protest.

What it is

Calves are reared in unnatural conditions, their movements deliberately restricted either in small pens or separate stalls and sometimes by tethering, in many cases deprived of light except at feeding times and even then given only artificial light, and fed on an unnatural diet including drugs. These methods are used to force quick growth and white meat. After lives of complete imprisonment the calves are slaughtered at the age of 12 weeks to give YOU CHEAP VEAL.

A growing evil

Similar intensified unnatural methods are now being extended to other animals. It is easy to see that unless public opinion calls a halt to this false progress NOW the day is very near when all our farm animals will be kept in factories tier upon tier.

We have proof of this in the frightening growth of the broiler chicken industry in this country. In 1960 the British public in their ignorance bought one hundred million broiler chickens. The industry confidently anticipates that the same public will purchase one hundred and thirty-five million chickens in 1961.

Image 3.5 Leaflet by the Crusade Against All Cruelty To Animals pushed through Harrison's letterbox around 1960 (image courtesy of Marlene Halverson)

Harrison's use of language, which strongly resembles the Quaker principle of living faith through action, is just as telling as her initial decision to campaign for animal welfare within the Quaker movement. However, despite forwarding the leaflet "to every Friends Meeting in the country,"[47] reactions were discouraging. Of a total of 20 replies, 18 said "there was enough suffering among humans without getting involved in animals."[48] For Ruth Harrison, this was not enough. Following the suggestion of a friend and supported by her husband,[49] she decided to write a comprehensive account to bear witness against intensive animal husbandry.[50]

[47] Spencer and Gerrel, "A rare breed at the factory farm".

[48] Spencer and Gerrel, "A rare breed at the factory farm".

[49] Spencer and Gerrel, "A rare breed at the factory farm".

[50] According to a later interview, Dex Harrison "backed [Ruth Harrison] to the hilt, although in the early days he was fairly detached about her work. Then one day he drove her to a research unit she wanted to look at. (…). The emotional impact on [Dex Harrison] was quite big." Kendall, "Ruth the Ruthless," 2; Dex Harrison also took many of the pictures, which appeared in *Animal Machines*.

Sixty years after the Crusade's leaflet landed in her letterbox, Harrison's decision to write *Animal Machines* remains remarkable. However, in contrast to heroic tales of her spontaneous emergence as a bestselling author,[51] surveying the years before 1960 reveals deep roots connecting *Animal Machines* and its author to the radical world of Edwardian reform. Her family background, Quakerism, training in the dramatic arts, and experiences within the FAU and CND made Harrison a perfect rebel waiting for a cause. Knowing this context is important. Harrison's experiences not only predisposed her to take up the topic of animal welfare but also made her target specific issues more than others. As dystopian sites of alleged animal suffering, environmental hazards, and moral degradation, 'factory farms' presented an ideal target for an activist concerned about modern technology's threat to peace, society, and the environment. This was true not only for Harrison but also for many of the other scientists, campaigners, and politicians gravitating towards animal welfare after 1945. As Part II shows, their rootedness in a similar cultural milieu and desire for a moral reordering of society facilitated personal friendships and would shape the trajectory of British farm animal welfare for decades to come.

[51] Webster, "Ruth Harrison," 6; Roscher, *Königreich*, 260; Ryder, *Animal Revolution*, 165.

PART II

Synthesis: The Post-war Landscape of Welfare Science and Activism (1945–1964)

Understanding the enormous impact of *Animal Machines* requires a detailed analysis of the scientific, social, and personal context in which it was written. Doing so shows that *Animal Machines* was about more than animal welfare and strongly shaped by Ruth Harrison's wider concerns about the ecological, moral, and health effects of contemporary food production. Beginning research for her book in 1961,[1] Harrison was writing during an auspicious time. On both sides of the Atlantic, ethologists were questioning supposed divides between animal and human cognition and anthropomorphic taboos associated with studying animal feelings (affective states). In Britain, the Universities Federation of Animal Welfare (UFAW) and ethologist William Homan Thorpe began translating research on evolutionary behaviour and stress into welfare guidelines. Their fusion of scientific and moral concerns fell on fertile political ground. With the first wave of Campaign for Nuclear Disarmament (CND) protests peaking, public concerns focused on environmental degradation, invisible health hazards, and wider moral threats to society. For Harrison and many others, the small but growing number of intensive farming operations seemed particularly problematic. As a Royal Academy of Dramatic Art (RADA)-trained dramatist, Harrison used the dystopia of the "factory farm" as a narrative centre piece to weave together

[1] FACT Files. Donald Broom [in the following DB], Box: Material for 'Animal Machines'.

popular concerns about intensive farming. Her ability to mobilise this dystopian "sociotechnical imaginary"[2] and stage a compelling romanticised alternative of animal welfare in a bucolic countryside would turn *Animal Machines* into an international bestseller and a defining document of post-war literary activism.

[2] Jasanoff and Sang-Hyun, "Sociotechnical Imaginaries," 189–196.

CHAPTER 4

Between Physiology and Psychology—Ethology and Animal Feelings

There is a long history of thinking about animals, their emotional states, and humans' duties towards the creatures in their care. For centuries, European thought was primarily influenced by the twin Christian concepts of humans' mastery over animals and animals' status as fellow creations of God. However, from the Early Modern period onwards, secularisation's gradual disenchantment of the world began to necessitate new concepts of animals' ontological, ethical, and legal status.[1] The modern recalibration of animals' societal status coincided with growing interest in animals' biological origins and behavioural characteristics. From the mid-twentieth century onwards, ethology emerged as an important new discipline explaining the roots and functioning of animal behaviour—and the welfare obligations resulting from it.

In nineteenth-century Britain, naturalists had already become intrigued by similarities between animal and human expression. Political affiliations exerted a strong influence on resulting theories about the natural world. In 1855, liberal utilitarian Herbert Spencer proposed that 'feelings'

[1] Keith Thomas, *Man and the natural world: Changing attitudes in England 1500-1800* (London: Penguin, 1991); Kean, *Animal Rights*, 13-38; see also Angela N.H. Creager and William C. Jordan, eds. *The animal-human boundary: historical perspectives* (Rochester: Rochester University Press, 2002), ix–xv.

C. Kirchhelle, *Bearing Witness*, Palgrave Studies in the History of Social Movements,
https://doi.org/10.1007/978-3-030-62792-8_4

alongside memory and reason enabled animals to adapt flexibly rather than just reflexively to their environment. All three characteristics played an important role in his Lamarckian theory of 'progressive' adaptive evolution.[2] Published one year after *The Descent of Man* in 1872, Darwin's *The Expression of Emotions in Man and Animals* provided detailed accounts of communicative behaviour in animals and humans. For the abolitionist Darwin, studying animal behaviour not only cast light on puzzling aspects of sexual selection but also served to highlight evolutionary continuities between species and the universality of humans.[3] Darwin's pupil, George Romanes, went on to explore links between animal and human consciousness and devised an evolutionary tree based on intelligence. Romanes' attempts to prove Darwinian evolution via comparative psychology led to the publication of *Animal Intelligence* in 1878[4] and *Mental Evolution of Animals* in 1883.[5]

The often-anecdotal way in which Darwin and Romanes mobilised evidence on animal behaviour proved controversial.[6] Although he also argued that animal cognition could be a positive evolutionary force, British psychologist and zoologist Conwy Lloyd Morgan became concerned about vague methodologies. Around 1900, he coined what became known as Morgan's Canon while studying relations between animal habit and instinct. Trying to put animal psychology and behavioural research on a more expert-based 'scientific' footing, Morgan stated that no animal activity should automatically be interpreted as a higher psychological process if it could also be explained by 'lower' processes of psychological evolution and development.[7] In other words, one should not by default attribute "higher" human concepts of rationality, purposiveness,

[2] David Fraser and Ian J.H. Duncan, "'Pleasures', 'Pains' and Animal Welfare: Toward a Natural History of Affect," *Animal Welfare* 7/4 (1998), 383–396; Piers J. Hale, *Political Descent. Malthus, Mutualism, and the Politics of Evolution in Victorian England* (Chicago: University of Chicago Press, 2014), 66-105.

[3] Charles Darwin, *The expression of emotions in animals and man* (*London: Murray*, 1872); Burkhardt, *Patterns of Behavior*, 72-75; John Sparks, *The Discovery of Animal Behavior* (London: William Collins & Co, 1982), 105-114; Adrian Desmond and James Moore, *Darwin's Sacred Cause. How a Hatred of Slavery Shaped Darwin's Views on Human Evolution* (Boston and New York: Houghton Mifflin Harcourt, 2009).

[4] John George Romanes, *Animal Intelligence* (New York: D. Appleton, [1878] 1884).

[5] John George Romanes, *Mental evolution in animals* (Kegan Paul, Trench, 1883).

[6] Sparks, *Discovery of Animal Behavior*, 118–126.

[7] Alan Costall, "Lloyd Morgan, and the Rise and Fall of Animal Psychology," *Society & Animals* 6/1 (1998), 16–29.

or affection to animal behaviour if this behaviour could also result from simple trial-and-error learning.[8]

While Morgan remained interested in animal psychology and cognition, other researchers attempted to show that many alleged instances of animal consciousness or voluntary behaviour were in fact due to physiological functions.[9] Mirroring a more general shift of biological research towards quantitative laboratory-based methods,[10] their work formed part of a growing backlash against 'unscientific' anthropomorphism. Around 1900, two researchers in particular laid the groundwork for a new era of mechanistic rather than introspective explanations of animal behaviour. In Russia, physiologist Ivan Petrovich Pavlov conducted iconic experiments on dogs, who salivated when hearing a bell associated with food (classical conditioning). In the US, psychologist Edward Lee Thorndike used puzzle boxes and mazes to measure cumulative animal learning (operant conditioning).[11] Animals' behaviour could be explained as resulting either from reflexes to certain environmental stimuli and motivational states or from controlling stimuli anchored in an individual's history. Focusing on animal cognition or affective states was unnecessary. Coined by US psychologist John Broadus Watson in 1913, a new school of 'behaviourists' tried to use predominantly mechanistic models to establish the behavioural disciplines as fully fledged natural sciences. In its most radical form, behaviourism—as promulgated by Burrhus Frederic Skinner in the US—completely discounted concepts of animal cognition and affective states that could not be tested experimentally.[12]

Although it dominated American research on mental states until the 1950s, behaviourism's authority was never absolute. On both sides of the Atlantic, competing fields like animal psychology or hybridised forms of behaviourism continued to evolve.[13] In 1917, German *Gestalt*

[8] Burkhardt, *Patterns of Behavior*, 76; Sparks, *Discovery of Animal Behavior*, 128.

[9] Sparks, *Discovery of Animal Behavior*, 129–141; 148–164.

[10] Robert E. Kohler, *Landscapes and Labscapes: Exploring the lab-field border in biology* (Chicago: University of Chicago Press, 2002), 1–19.

[11] Sparks, *Discovery of Animal Behavior*, pp . 148–155.

[12] Katja Guenther, "Monkeys, Mirrors, And Me: Gordon Gallup And The Study of Self-Recognition," *Journal of the History of the Behavioural Sciences* 53/1 (2017), 11–14; Sparks, *Discovery of Animal Behavior*, 155–164; instances of insight learning had also been described by sociologist L.T. Hobhouse in Manchester.

[13] Ingo Brigandt, "The instinct concept of the early Konrad Lorenz," *Journal of the History of Biology* 38/3 (2005), 575–576; Guenther, "Monkeys," 12–14.

psychologist Wolfgang Köhler highlighted instances of insight learning by chimpanzees on Tenerife, whose ability to use poles and stack boxes to reach bananas seemed not to stem from cumulative trial-and-error learning but from the internal realisation of a solution.[14] In the US, psychologist William McDougall posited that animal instincts were more than mere reflexes and could be informed, motivated, and modulated by subjective experiences (emotions).[15] At Cornell, psychobiologist Howard Liddell pioneered the comparative study of neuroses in animals and humans.[16]

Another group of researchers used approaches from field studies and comparative psychology to study innate and acquired animal behaviours in relatively 'natural' conditions.[17] Applying evolutionary theory to account for behavioural continuities among related species, so-called ethologists were interested both in the biological usefulness of instinctive behaviour and in some animals' ability to adaptively modify behaviour and acquire new knowledge via learning. Building on earlier work by naturalists and psychologists like Oskar Heinroth and Edmund Selous, their findings led ethologists to challenge purely mechanistic concepts of behaviour.[18]

Between the 1910s and 1930s, zoologist Julian Huxley conducted pioneering research on the ritualised behaviour of birds, including great crested grebes. Huxley linked observed behavioural patterns to Darwinian evolutionary theory but challenged parts of Darwin's theory of sexual selection by noting that ritualised behaviour linked to courtship continued after pair formation and reproduction. He also noted that long-term sexual selection for certain traits might not always be beneficial for non-reproductive fitness.[19]

However, it was in continental Europe that ethology would truly take off. In 1920s' Austria, biologist Karl von Frisch described colour perception, orientation, dance-like communication, and dialects among honeybees.[20] A few years later, his contemporary Konrad Lorenz conducted

[14] Sparks, *Discovery of Animal Behavior*, 166.

[15] Fraser and Duncan, "Pleasures".

[16] Robert G. W. Kirk and Edmund Ramsden, "Working across species down on the farm: Howard S. Liddell and the development of comparative psychopathology, c. 1923–1962," *History and Philosophy of the Life Sciences* 40/24 (2018); see also Burkhardt, *Patterns of Behavior*, 17–68.

[17] Brigandt, "The instinct concept," 593–594.

[18] Burkhardt, *Patterns of Behavior*, 78–103; Brigandt, "The instinct concept," 573–574.

[19] Burkhardt, *Patterns of Behavior*, 103–126.

[20] Sparks, *Discovery of Animal Behavior*, 180–189.

ground-breaking work on imprinting by young birds. Breaking with goal-focused psychological explanations of behaviour, Lorenz was convinced that physiological rather than psychological methods should be used to explain animal behaviour in a scientific manner. According to Lorenzian ethology, animal behaviour had evolved over time due to Darwinian selection. Like a film sequence, researchers should break down complex behaviours into individual components (e.g. innate motor patterns) and analyse these components in relation to the wider physiological sequence. By breaking down behaviour like egg rolling by geese, it could be compared taxonomically among different animal species and its evolutionary origins could be reconstructed phylogenetically.[21]

According to Lorenz's 'hydraulic' model of behaviour, animals were primed to carry out innate behaviour patterns that were triggered by environmental stimuli. Once a routine was finished, it would stimulate the next appropriate behaviour. Given the internal build-up of sufficient 'action-specific' behavioural energy, animals would first search for the appropriate stimulus (appetite behaviour) for a behaviour and eventually engage in this behaviour even in the absence of an appropriate stimulus (vacuum activities). Animals likely had no awareness of the purpose of a behaviour, and their lived experience would not change basic innate instincts. While the goal-directedness of observed behaviours could be explained in evolutionary terms, researchers in the field should not confuse these (ultimate) causes with the (proximate) physiological and environmental triggers stimulating it.[22]

Dutch biologist Nikolaas (Nikko) Tinbergen was also sceptical of 'subjective' animal psychology and laboratory-based 'mechanistic' behaviourist research. Characterised by historian Richard Burkhardt Jr as a hunter, he preferred research on 'wild' animals like digger wasps and herring gulls. Tinbergen had started researching animal behaviour during the 1930s and collaborated with Lorenz in Austria. Spending the wartime years in a Nazi prisoner camp, Tinbergen moved to Oxford in 1949. Although he initially shared Lorenz's focus on physiological mechanisms of behaviour, criticism from US psychologist Daniel Lehrman and British ethologist Robert Hinde made Tinbergen abandon notions of unchangeable innate behaviour as well as his own—more elaborate—cascade model of hydraulic behaviour. From the early 1950s onwards, Tinbergen instead emphasised

[21] Brigandt, "The instinct concept," 571–608, particularly 576–577 and 602–604.

[22] Brigandt, "The instinct concept," 576–578; Burkhardt, *Patterns of Behavior*, 127–186.

comparative field studies of animal behaviour and analyses of behaviours' biological utility (survival utility) in complex ecological settings.[23] Drawing on work by Julian Huxley, Ernst Mayr, and Konrad Lorenz, he developed four complementary analytical approaches to interpret animal behaviour. Definitively set out in 1963, Tinbergen's "four questions" provide the now classic definition of ethology and integrate a proximate physiological analysis of behaviour focusing on the (1) immediate causation of behaviour (mechanism) and its (2) ontogeny (development) with an ultimate analysis of behaviour focusing on the (3) function of a behaviour (evolutionary adaptation) and its (4) phylogeny (evolution). Answering the four questions may yield different answers, but answers will not contradict answers for the other questions.[24]

By the late 1950s, European ethology was displacing laboratory-based behaviourist models with more complex environmentally situated and evolutionarily rooted explanations of animal behaviour. The field's growing prominence was internationally recognised in 1973 when Tinbergen and Lorenz were jointly awarded the Nobel Prize in Physiology or Medicine with Karl von Frisch for their "discoveries concerning organization and elicitation of individual and social behaviour patterns."[25] However, both Tinbergen and Lorenz consistently shied away from engaging with animal behaviour beyond directly observable traits. Fearing the Damocles sword of anthropomorphism and—in the case of Lorenz—committed to models of immutable innate behaviour, many continental ethologists avoided normative discussions of animal cognition, feelings (affective states), or welfare in laboratories or on farms.[26] Writing to Julian Huxley in 1959, Niko Tinbergen noted:

> I am willing to admit that members of the same species can be supposed to feel roughly the same, but I think it is just futile even to try and argue [that one can understand how other species feel]—it is just not arguable. I will just willingly concede that we may suppose they feel the same. I cannot see how one can ever know whether one experiences what another man, another

[23] Burkhardt, *Patterns of Behavior*, 187–230; 362–368; 371–373; 378; 382–387; 399–402; 408–434.

[24] Richard W. Burkhardt Jr, "Tribute to Tinbergen: Putting Niko Tinbergen's 'Four Questions' in Historical Context," *Ethology* 120 (2014), 215–223.

[25] Burkhardt, *Patterns of Behavior*, 447–459.

[26] Burkhardt, *Patterns of Behavior*, 434–436.

mammal, another vertebrate, another animal, another organism feels. That for me ends the matter—as long as we practice science.[27]

Tinbergen's compunctions were not shared by British ethologists. Similar to prominent scientists' rise as public intellectuals in other Western countries,[28] the post-war years saw British behavioural researchers and biologists promote science's role as a progressive socio-moral force.[29] Although Lorenz and Tinbergen also engaged with contemporary political debates,[30] their British counterparts believed that understanding animal cognition and feelings was key to developing a more humane science that would improve both animal welfare and society's moral status. This thinking had a long tradition. By the early twentieth century, the antivivisection movement had sensitised generations of British researchers to ethical issues surrounding the treatment of animals and the societal embeddedness of their work.[31] According to Robert Kirk, there was "broad agreement that the resulting Cruelty to Animals Act (1876) was positive for science and society alike."[32]

In 1926, leading zoologists, veterinarians, and scientists founded the Universities Federation for Animal Welfare (UFAW). Representing a 'distinct amalgamation of pragmatic science and humane moral values,'[33] the UFAW tried to develop scientific solutions for animal welfare. Although it distanced itself from the 'anti-scientific sentimentality' of antivivisectionists, senior members like Charles Hume saw anthropomorphism 'as a means to understand animal experience.'[34] Drawing on traditions of

[27] Rice University Archives, Julian Sorell Huxley Papers, Box 21, Folder 1, Nikolaas Tinbergen to Julian Huxley, 21.01.1959, 1.

[28] Cathryn Carson, "Bildung als Konsumgut: Physik in der westdeutschen Nachkriegskultur," in Dieter Hoffmann (ed.), *Physik im Nachkriegsdeutschland* (Frankfurt: Harri Deutsch, 2003), 73–85; Cathryn Carson, "Science as instrumental reason: Heidegger, Habermas, Heisenberg," *Cont Philos Rev* 42 (2010), 483–509; David C. Engermann, "Social science in the Cold War," *Isis* 101/2 (2010), 395–399.

[29] See for example Burkhardt on Alistair Hardy in Oxford, Burkhardt, *Patterns of Behavior*, 333.

[30] Chloe Silverman, "'Birdwatching and baby-watching': Niko and Elisabeth Tinbergen's ethological approach to autism," *History of Psychiatry* 21/2 (2010), 176–189; Burkhardt, *Patterns of Behavior*, 454–456.

[31] Kean, *Animal Rights*, 96–112.

[32] Robert G.W. Kirk, "Science and humanity".

[33] Kirk, "Science and humanity".

[34] Kirk, "Science and humanity".

Christian social reform and evolutionary theories, Hume and others endorsed contemporary notions of kinship between animals and humans and developed a synthesist brand of "scientific humanism."[35] For them, science was best positioned to morally improve society by finding rational ways to alleviate human and animal suffering. However, only a humane science could produce a humane society. In contrast to unsystematic earlier attempts to reduce cruelty, welfare emerged as a positive systematic concept, which could be measured, and regulated.[36]

The Second World War catalysed UFAW efforts. Reports about barbaric Nazi experiments, new killing technologies, and Winston Churchill's warnings about "perverted science" leading into the "abyss of a new Dark Age"[37] challenged progressivist doctrines of science as a force for moral progress. While the UFAW had previously addressed a plethora of issues ranging from the electrocution of slaughter animals to vermin control, the organisation increasingly focused on the humane treatment of animals used to produce scientific knowledge in the laboratory.[38] Initially targeting technicians rather than scientists, the UFAW's engagement with laboratory practice was facilitated by new ethological interest in animal behaviour and contemporary calls for a more standardised supply of research animals. The emerging concept of stress was particularly important and enabled UFAW researchers like pharmacologist Michael Chance to present 'well-being' as a moral concern and a legitimate area of practical scientific inquiry. Using ethological methods to identify 'normal' species-specific behaviour of lab animals, Chance showed that different forms of social behaviour altered animals' reaction to pharmaceutical drugs. Stressed animals in particular produced unreliable results. Stress could also be quantified. This opened the door to assessing well-being in different laboratory systems, establishing best practice guidelines, and warding off accusations of unscientific anthropomorphism.[39]

[35] Roger Smith, "Biology and values in interwar Britain: CS Sherrington, Julian Huxley and the vision of progress," *Past & Present* 178 (2003), 235.

[36] Kirk, "Science and humanity"; Kirk, "Recovering the Principles of Humane Experimental Technique," 624–632.

[37] Winston Churchill, "Their Finest Hour", 18.06.1940, House of Commons, https://winstonchurchill.org/resources/speeches/1940-the-finest-hour/their-finest-hour/ [13.05.2020].

[38] Kirk, "Recovering the Principles of Humane Experimental Technique," 628–632.

[39] Robert G.W. Kirk, "Between the Clinic and the Laboratory: Ethology and Pharmacology in the Work of Michael Robin Alexander Chance, c. 1946–1964," *Medical History* 53 (2009), 513–536.

In Oxford, zoologist William Moy Stratton Russell subsequently used indicators like breeding productivity and animals' behaviour towards experimenters to create a welfare scale ranging from well-being to distress. Russell argued that good—and thus humane—research should aim to enhance well-being and reduce stress. In 1959, he co-authored *The Principles of Humane Experimental Technique* with Rex Burch. The publication laid out the influential 3Rs (replacement, reduction, and refinement) as a new ethical gold standard for the design of animal-dependent science. According to Robert Kirk, *The Principles* marked a late highpoint of the foregrounding of science's moral mission within society.[40] To achieve this moral mission during a time of growing concerns about humans' technological alienation from 'nature' and ethics, science had to actively incorporate humane values into its own practice.

UFAW members' synthesist view of science, society, and ethics was shared by ethologist William Homan Thorpe. Working at Cambridge since 1932, Thorpe had begun his career as a neo-Lamarckian believing in the inheritance of acquired characteristics and researching insect behaviour and control. By the late 1930s, he had abandoned Lamarckianism in favour of Darwinian models of evolution and become interested in behavioural preferences' role in driving speciation ("Baldwin Effect"). After reading the work of Konrad Lorenz, Thorpe shifted his research to studying instinct and learning in higher animals.[41]

As described by Gregory Radick and Neil Gillespie, this turn to ethology was influenced by Thorpe's religious beliefs and increasingly outspoken support of Quakerism. Thorpe had started attending meetings of the Society of Friends around 1930 and formally converted to Quakerism after 1945. A conservationist and self-described "amateur philosopher-theologian,"[42] Thorpe saw ethology as a way to reconcile his Christian beliefs with the modern synthesis of natural selection and Mendelian genetics by highlighting the "wholeness, of man and nature, and of the relation of both to the divine."[43] Thorpe's natural theology rested on the

[40] Kirk, "*Recovering the Principles of Humane Experimental Technique*," 640–641.

[41] Burkhardt, *Patterns of Behavior*, 337–341; Hinde, "William Homan Thorpe," 621–639; Gregory Radick, "Animal agency in the age of the Modern Synthesis," 41–45.

[42] Quoted according to: Hinde, "William Homan Thorpe," 626, see also: 625, 630–633; Oral History Interview Donald Broom (04.07.2014); Gillespie, "The interface of natural theology and science," 27.

[43] Gillespie, "The interface of natural theology and science," 4.

concept of 'creative' evolution and was influenced by sociologist Leonard Hobhouse's theory of evolution as a mind-expanding process, which had become self-conscious in humans.[44] According to Thorpe, human consciousness was the result of natural selection and emergence—an evolutionary event that could not be fully predicted by an organism's phylogeny (evolution) or ontogeny (development). The emergence of consciousness was not mechanistic but 'creative' and its origins were purposeful. In addition to opening the door for divine design, postulating that consciousness was an emergent event with evolutionary power allowed Thorpe to reconcile humans' animal origins with a view of their cognitive and spiritual uniqueness.[45] Thorpe's theory of cognitive and behavioural evolution made him conceive of ethology as a somatic (bodily) and mental science.[46] It also necessitated seeing animals as something more than mindless machines, whose kinship with humans necessitated empathy.

Thorpian thinking influenced both British ethology and animal welfare politics. Having potentially met the young Ruth Harrison at Cambridge's Quaker Meeting House or in his role as conscientious objector and chairman of the local Pacifists' Service Bureau (see Chap. 3), Thorpe played a key role in bringing continental ethology to Britain after 1945. As the founding editor of *Behaviour*, he not only lobbied for a post for Lorenz in the UK but also helped engineer a reconciliation between Lorenz and Tinbergen—the former an erstwhile Nazi supporter and the latter a Nazi victim.[47] In Cambridge, Thorpe established and became the director of the Ornithological Field Station at Madingley, where he and his pupils researched the role of instinct, purposive behaviour, and insight learning in the development of chaffinch song. His research on wild birds and birds reared in auditory isolation revealed that learning occurred in two stages, with chaffinches learning first what to sing and then how to sing it. Additional work focused on breaking down instances of insight learning by birds and analysing the extent to which other species also imprinted on parents and whether this process was reversible.[48] Thorpe's findings not only challenged Lorenz's theory of immutable instincts and Tinbergen's

[44] Radick, "Animal agency in the age of the Modern Synthesis," 46–47.

[45] Gillespie, "Interface of Natural Theology and Science," 4–7; Radick, "Animal agency in the age of the Modern Synthesis," 35–56.

[46] Gillespie, "Interface of Natural Theology," 21–29.

[47] Hinde, "William Homan Thorpe," 625 and 627; Burkhardt, *Patterns of Behavior*, 341–345.

[48] Hinde, "William Homan Thorpe," 628–629; Burkhardt, *Patterns of Behavior*, 340–345; Gillespie, "Interface of Natural Theology," 24–27, 35.

agnosticism regarding animals' affective states but also helped keep British ethology open to questions of welfare. If animals were self-aware and could think and learn, then frustrating their 'normal' behavioural impulses and needs—a concept developed by Thorpe's student and colleague Robert Hinde[49]—could be emotionally and mentally detrimental even if it did not cause physical harm.

Thorpe's synthesist focus on animals' affective states, cognitive evolution, and human morality was shared by Julian Huxley—albeit for entirely secular reasons. Since starting his research on bird behaviour in 1907, Huxley had credited animals with mental states and repeatedly stated that it was possible to know what these states were. Interpreting evolution as a progressive force for the good of a species, Huxley came to the conclusion that biological explanations of behaviour could not be reduced to aspects of sexual selection.[50] After ending his university career in the mid-1920s, Huxley became a public intellectual. In 1927, his bestselling *The Science of Life*, co-authored with H.G. and G.P. Wells, fused ecological, behavioural, and evolutionary concepts to argue for a synthesis between Darwinian evolution and Mendelian genetics. During the 1930s, Huxley intensified work with geneticists, mathematicians, and population biologists and coined the term "modern synthesis" in 1942.[51] After a brief stint as the first director of the United Nations Educational, Scientific and Cultural Organization (UNESCO, 1946–1948), the 1950s saw Huxley re-emerge as a prominent British commentator on evolutionary theory, ethological research, wildlife conservation, and societal mores.

Similar to Thorpe, Huxley was particularly interested in the role that supposedly emergent evolutionary traits like language or cognition could play in shaping further evolution (Baldwin Effect).[52] However, as a staunch atheist, Huxley disagreed with Thorpe on the causes of this emergence. Whereas Thorpe thought of the mind as the result of 'creative' emergence, Huxley thought of it as resulting from a secular but progressive evolutionary process. Writing to Thorpe in 1963, he noted:

[49] Donald M. Broom, "World Impact of ISAE: past and future," in Jennifer Brown, Yolande Seddon and Michael Appleby (eds.), *Animals and Us—50 years and more of applied ethology* (Wageningen: Wageningen Academic Publishers, 2016), 270.

[50] Burkhardt, *Patterns of Behavior*, 107–108, 110, 121–122.

[51] Burkhardt, *Patterns of Behavior*, 119–125.

[52] Radick, "Animal agency in the age of the Modern Synthesis," 56.

> you and I start from different premises. You assume the existence of absolutes and of some sort of super-person, or supernatural being, whereas I start from just the opposite angle … absolutes seem to me to be either the product of logical abstraction taken to an unreal extreme, or else … the result of all-or-nothing processes in our minds…[53]

Rather than "reifying" the existence of the mind, Huxley advocated focusing on individual physiological components of cognition and promoted the term psychometabolism. Thorpe disagreed emphatically: "Of course I am reifying mind—that is the whole point. If I didn't think that you had an entity such as mind I shouldn't pay any attention at all to what you are saying." According to Thorpe, secularists like Huxley were also assuming evolutionary absolutes but not admitting it:

> I believe I could literally find a hundred statements or sentences in your various written works which imply that you do assume the existence of absolutes. I don't want to start playing the game of the mote and the beam … but I do think some of these fundamental ideas at the basis of scientific thinking want to be brought out into the open, and I find personally that any of those who call themselves 'humanists' are wildly inconsistent just because they fail to do this.[54]

This criticism was not unfair. Whereas Thorpe interpreted the emergent properties of the human mind as a sign of 'creative' divine evolution, which enabled humans to strive for spiritual improvement, Huxley conceived of cognition as the emergent result of precarious but ultimately progressive evolutionary forces.[55] As an advocate of 'transhumanism,' Huxley hoped that humans' cognitive abilities would allow them to take charge of their evolutionary future in the form of a new religion of evolution, which included eugenics and birth control.[56] A better ethological analysis of the evolutionary roots of human behaviour—including

[53] Rice University Archives, Julian Sorell Huxley Papers, Box 34, Folder 1, Julian Huxley to William Homan Thorpe, 10.01.1963, 1.

[54] Rice University Archives, Julian Sorell Huxley Papers, Box 34, Folder 1, William Homan Thorpe to Julian Huxley, 26.01.1963, 1.

[55] Burkhardt, *Patterns of Behavior*, 122–125.

[56] Julian Huxley, "Transhumanism," *Journal of Humanistic Psychology* 8/1 (1968), 73–76; Alison Bashford, "Julian Huxley's Transhumanism," in Marius Turda (ed.), *Crafting Humans: From Genesis to Eugenics and Beyond* (Göttingen and Taipei, V&R Uni Press/National Taiwan University Press, 2013), 153–162.

affective states and cognition—was crucial to this endeavour. This put him at odds with Tinbergen's insistence that animal feelings were outside the purview of ethology. Writing to Tinbergen in 1965, Huxley noted:

> As regards subjective phenomena, it still seems to me that, just as one deduces that other human beings have subjective experiences of different sorts in relation to different circumstances, so we can and must deduce that different types of situations in animals are accompanied by different subjective states. ... Total exclusive reliance on subjective approach is of course a very serious obstacle to progress in human psychology—but so is an uncritical materialist-behaviourist approach ... I feel that ethology could and should become the basis for a real science of human psychology, and if so, it has to take account of subjective phenomena which by definition are involved in psychology![57]

The years after 1945 thus saw the synthesist orientation of British research on animal behaviour extend ethology's remit to encompass animals' cognitive and affective states. Breaking with continental ethologists and driven by senior British researchers' theist and non-theist beliefs, this development in turn opened the door for active scientific engagement with the politics of animal welfare. It also created significant synergies with the moral agendas of emerging activists like Ruth Harrison, who often came from similar cultural and religious *milieus*. As a result of these synergies, the intensive farm would emerge as a crucible for activist and scientific thinking about animal welfare and the moral status of the humans producing these animals.

[57] Rice University, Julian Sorrell Huxley Archives, Box 38, Folder 5, Julian Huxley to Nikolaas Tinbergen, 29.03.1965.

CHAPTER 5

Ideals and Intensification: Welfare Campaigns in a Nation of Animal Lovers

Campaigning for animal protection had an illustrious history in Britain. Starting in the late eighteenth century, an increasing number of Britons had called for the improved treatment of animals. Founded in 1824, the RSPCA was the first organised body for animal protection and had influential supporters in parliament and society. During the second half of the nineteenth century, antivivisectionist and anti-cruelty campaigns commanded considerable public support.[1] As numerous authors have emphasised, the prominence of British debates about the proper treatment of animals was exceptional.[2] It was also reflected in a series of laws centring on animal protection. Motivated by humanitarian and disease concerns, a series of Parliamentary Acts had introduced measures to protect animals in transit from the 1870s onwards. In 1911, the Protection of Animals Act consolidated existing rules. Restricted to public spaces, the 1911 Act was intended to prevent wilful physical cruelty to animals. However, enforcement of the 1911 Act and of the enhanced provisions for animal slaughter (1954 and 1958) was lackluster. Meanwhile, concerned RSPCA inspectors

[1] Kean, *Animal Rights*, 35–135.
[2] See, for example, Kean, *Animal Rights*, 1–12; and Roscher, *Königreich*, 11–15.

C. Kirchhelle, *Bearing Witness*, Palgrave Studies in the History of Social Movements,
https://doi.org/10.1007/978-3-030-62792-8_5

were not allowed to enter premises against an owner's will.[3] As a consequence, relatively few successful convictions resulted from the Act.[4]

Despite such legislative gaps and criticism by humanitarian campaigners like Henry Salt (see Chap. 2), the trope of being a 'Nation of Animal Lovers' became increasingly popular in Britain. As described by Hilda Kean, this alleged national trait was reinforced by British charities and officials during the two world wars, when campaigns emphasised civilised British compassion as opposed to German cruelty. 'British' character traits were also superimposed onto national animal breeds. Although several hundreds of thousands of cats, dogs, and other pets were euthanised at the outset of the war in 1939, most owners decided to spare their pets' lives. The following years saw campaigns emphasise either Britons' compassionate treatment of animals or British animals' heroism in the face of the mechanised Nazi onslaught.[5]

Following the war, the 'Nation of Animal Lovers' trope strengthened the social and political standing of animal charities like the RSPCA. However, in what can be described as "organisational capture,"[6] the Society's establishment status and close ties to Britain's elite also decreased its willingness to confront authorities. Similar to arrangements between agricultural officials and the National Farmers Union (NFU), many aspects of RSPCA animal protection were quasi-corporatist:[7] the Society was granted limited

[3] Woods, "From Cruelty to Welfare," 14–15; Neil Pemberton and Michael Worboys. *Rabies in Britain: Dogs, Disease and Culture, 1830–2000* (London: Palgrave Macmillan, 2006), 40–101; Webster, "Ruth Harrison—Tribute," 5.

[4] The British National Archives [in the following TNA] MAF 260/351 Part II - Protection of Animals Act, enclosed in, Background Notes and Possible Supplementary Questions, enclosed in: Minute ES Virgo to Mr Hutchison (18.03.1964), 1.

[5] Kean, *Animal Rights*, 166–179; 191–197; Roscher, *Königreich*, 242–243; Hilda Kean, *The Great Cat and Dog Massacre. The Real Story of World War Two's Unknown Tragedy* (Chicago: University of Chicago Press, 2017), Chapters three, six, and seven; for a broader discussion of animals and nationalist character assignations see: Harriet Ritvo, *The Animal Estate. The English and Other Creatures in Victorian England* (Cambridge MA: Harvard University Press, 1987) and Sandra Swart, "The other citizens: Nationalism and animals," in Hilda Kean and Philip Howell (eds.), *The Routledge Companion to Animal-Human History* (London: Routledge, 2018), 31–52.

[6] John Abraham, *Science, Politics and the Pharmaceutical Industry. Controversy and Bias in Drug Regulation* (London and New York: Routledge, 1995), 22–23.

[7] Michael Winter, "Corporatism and agriculture in the UK: the case of the milk marketing board," *Sociologia Ruralis* 24/2 (1984), 106–119; Graham Cox, Philip Lowe, and Michael Winter, "From State Direction to Self-Regulation: The Historical Development of Corporatism in British Agriculture," *Policy and Politics* 14/4 (1986), 475–490; Michael

powers like inspection rights and was regularly consulted when it came to developing new regulations. In return, RSPCA leaders were expected to create acceptance for or moderate criticism of official politics. Traditional forms of lobbying via parliamentary channels, letters to the *Times*, and backroom influencing were acceptable—instigating public mass-protests or using new forms of direct action was not.[8]

During the 1950s and 1960s, two issues increasingly strained corporatist welfare arrangements: the first was already entrenched and rising RSPCA grassroots protest against 'field sports' like hare coursing, stag carting, and fox hunting with hounds—pastimes enjoyed by the RSPCA's royal patrons and some of its elite members. The second was the perceived rise of the 'factory farm.'[9]

Starting in the interwar period, some British farmers had attempted to compete with Danish bacon imports by experimenting with new kinds of intensive indoor pig production. However, a combination of foreign competition, disease pressure, and *laissez-faire* policies had made many endeavours fail.[10] Despite growing government involvement in agricultural production, the Second World War saw a significant decline in British animal numbers. The official wartime emphasis on caloric output meant that farmers focused on plant instead of animal production. With the exception of the dairy sector, British animal production fell dramatically.[11] After 1945, the situation did not improve immediately. Harvest failures, wartime destruction, the sudden termination of the American Lend Lease agreement, and enforced sterling-dollar convertibility left Britain chronically short of cash and made feedstuff imports undesirable. Bogged down by its military commitments in Europe and in its colonies, Britain prolonged production controls and rationing until 1954.[12]

Following the end of rationing, things changed rapidly. Having profited from wartime price guarantees and having been encouraged to expand

Winter, *Rural Politics: Policies for Agriculture, Forestry and the Environment* (London: Routledge, 1996), 3, 19–21, 100–103, 115.

[8] Roscher, *Königreich*, 242–45.

[9] Roscher, *Königreich*, 231–242, 290–291, 294–297.

[10] Abigail Woods, "Rethinking the History of Modern Agriculture: British Pig Production, C. 1910–65," *Twentieth Century British History* 23/2 (2012), 168.

[11] John Martin, *The Development of Modern Agriculture. British Farming since 1931* (London et al.: Macmillan & St. Martin's Press, 2000), 6–8; 10; 23; 38; 51; 54.

[12] Ina Zweiniger-Bargielowska, *Austerity in Britain: Rationing, Control and Consumption 1939–1955* (Oxford and New York: Oxford University Press, 2002), 73.

production by the 1947 and 1957 Agriculture Acts, British farmers invested heavily in new technologies.[13] Consumers' hunger for meat also fuelled an expansion of livestock production.[14] Whereas Britain produced 762,000 tonnes of meat in 1947, total meat production more than doubled to 1,713,000 tonnes in 1960.[15] Described by Abigail Woods, many initial production gains in pig farming were achieved within existing outdoor or hybrid indoor-outdoor systems.[16] However, in the poultry sector, a growing number of producers adopted new intensive indoor facilities, designed to guarantee the year-round production of animals via optimised environments and feed regimes. Declining feedstuff costs, consumer demand, and improved disease control facilitated a rapid increase in production and concentration processes within industry.[17] By the early 1960s, 35 per cent of British laying stock were located in battery cages and 50 per cent were produced in indoor deep litter houses. Two-thirds of broiler chickens were kept indoors in units of more than 20,000 birds.[18] Inspired by the poultry industry, cheap grain, and continental success stories, new intensive production methods were also trialled in pig and calf production.[19]

Although Karen Sayer and Abigail Woods have shown that 'factory farming' was far from ubiquitous around 1960,[20] the changes wrought by existing intensive animal production facilities and the likelihood of further intensification appeared dramatic to many contemporaries. In stark contrast to popular images of bucolic countryside pastures, animals in intensive

[13] B. A. Holderness, *British Agriculture since 1945* (Manchester: Manchester University Press, 1985), 12–16; 21.

[14] H. J. H. Macfie and Herbert L. Meiselman, *Food Choice Acceptance and Consumption* (London: Blackie Academic & Professional, 1996), 377.

[15] Europe: Meat Output Statistics, in: "International Historical Statistics" (London: Palgrave Macmillan, April 2013).

[16] Woods, "Rethinking".

[17] Andrew Godley and Bridget Williams, "Democratizing luxury and the contentious "invention of the technological chicken" in Britain," *Business History Review* 83/2 (2009), 267–290; Andrew Godley, "The emergence of agribusiness in Europe and the development of the Western European broiler chicken industry, 1945 to 1973," *Agricultural History Review* 62/2 (2014), 315–336; Alessandra Tessari and Andrew Godley, "Made in Italy. Made in Britain. Quality, brands and innovation in the European poultry market, 1950–80," *Business History* 56/7 (2014), 1057–1083.

[18] Woods, "Cruelty to Welfare," 16.

[19] Woods, "Cruelty to Welfare," 16; Woods, "Rethinking".

[20] Sayer, "Animal Machines," 482–83, Woods, "Rethinking".

operations were confined in high-density settings and bred and fattened for maximum productivity. While some commentators hoped that efficient production would prevent Malthusian scenarios of global overpopulation and political instability,[21] others began to question the wider logic of intensification. Popular criticism of intensive animal production had three interwoven strands: (1) a first strand focused on potential personal and environmental health hazards resulting from new intensive methods and technologies like antibiotics and pesticides; (2) a second strand focused on the physical and emotional welfare of intensively produced animals; (3) a third strand highlighted the 'alien' nature of 'factory farms' and its threat to British values and 'the countryside.' Over time, the emerging dystopian "sociotechnical imaginary"[22] of the 'factory farm' would not only mobilise significant public and consumer protest but also lead to significant changes in animal policymaking (Chaps. 7–9) and open the door for a new discipline of welfare science (Chap. 10).

In the case of health and environmental concerns, the interwar period had triggered a gradual integration and formalisation of British 'anti-chemical,' vegetarian, and other dietary reform movements. Inspired by developments in continental Europe as well as by work on composting and 'natural' diets by agricultural scientist Albert Howard and physician Robert McCarrison, a small group of British landowners and consumers began to experiment with 'non-artificial' forms of nutrition and food production. Founded in 1946 by Eve Balfour, niece of former Prime Minister Arthur Balfour, the Soil Association dedicated itself to developing non-intensive forms of organic agriculture.[23] Although the initial reach of this elite organisation was limited, celebrity endorsements, food scares, and 1950s' books like Franklin Bicknell's *The English Complaint* (1952)[24] and

[21] G. R. H. Nugent, "The Twentieth-Century Hen", *Times*, 30.07.1951, 5; "Pigs Fattened By Antibiotics", *Times*, 01.12.1952, 3; Anthony Lisle, "Untouched by Hand", *Farmers Weekly*, 06.07.1962, 99; Kirchhelle, *Pyrrhic Progress*, 79–80.

[22] Jasanoff and Sang-Hyun, "Sociotechnical Imaginaries," 189–196.

[23] R. More-Colyer, "Towards 'Mother Earth': Jorian Jenks, Organicism, the Right and the British Union of Fascists," *Journal of Contemporary History* (2004), 353–371; Philip Conford and Patrick Holden, "The Soil Association," in William Lockeretz (ed.), *Organic Farming: An International History* (Wallingford, UK: CABI, 2001), 187–192; Philip Conford, *The Origins of the Organic Movement* (Edinburgh: Floris, 2001), 83–92, 146–151.

[24] Franklin Bicknell, *The English Complaint or Your Fatigue and its Cure* (London: William Heinemann, 1952).

Doris Grant's *Housewives Beware* (1958)[25] spread the message of healthy 'natural' food and dangerous intensive production methods.[26]

Concerns about intensive farming's health hazards reached a peak around 1960. In that year, future Pulitzer Prize winner William Longgood published *The Poisons in Your Food*.[27] In his book, Longgood warned consumers about the invisible toxins and carcinogens contaminating their food. According to Longgood, intensive technologies in plant and animal production had turned consumers' shopping carts into a toxic ensemble. Despite being attacked as "an all-time high in 'bloodthirsty pen-pushing,'"[28] Longgood's 1960 publication became a bestseller and coincided with the opening of the British Soil Association's first Wholefood store for organic produce in London.[29] Two years later, American biologist Rachel Carson heightened concerns about the fallout of the 'chemical revolution' in her iconic *Silent Spring*. Carson's book—which appeared in Britain in 1963—fused long-standing concerns about chemical pesticides' and insecticides' environmental impact with more intimate concerns about personal health. Now often remembered for its powerful attack on DDT, *Silent Spring* was wary of an overall increase in agricultural and environmental chemical use. For Carson, intensive production systems' reliance on chemical helpers was facilitating the unchecked spread of carcinogens and toxic substances into citizens' environment, food, and bodies.[30]

Carson and Longgood's views were part of a rapidly expanding sea of environmentalist-inspired health warnings. In 1960, British physician Franklin Bicknell rehashed earlier warnings about unnatural production methods in his *Chemicals in Food and in Farm Produce*.[31] Three months ahead of *Silent Spring*, the American anarchist and libertarian socialist Murray Bookchin published similar warnings about modern agriculture's

[25] Doris Grant, *Housewives Beware* (London: Faber and Faber, 1958).

[26] Kirchhelle, *Pyrrhic Progress*, 80–83.

[27] William Longgood, *The Poisons in Your Food* (New York: Simon and Schuster, 1960).

[28] William J. Darby, "Review, the Poisons in Your Food by William Longgood," *Science*, 131/3405 (1960), 979.

[29] Craig Sams, "Introduction," in Simon Wright (ed), *Handbook of Organic Food Processing and Production* (Dordrecht: Springer Science + Business Media 1994), 12.

[30] Rachel Carson, *Silent Spring* (New York: Houghton Mifflin, 1962); organisations like the Audubon Society had been warning about the use of DDT for years; Simon, *Ddt. Kulturgeschichte Einer Chemischen Verbindung* (Basel: Christian Merian Verlag, 1999), 14–21, Edmund Russell, *War and Nature. Fighting Humans and Insects with Chemicals from World War I to Silent Spring* (Cambridge and New York: Cambridge University Press, 2001), 204–23.

[31] Bicknell, *Chemicals in Food*.

environmental and health hazards under the pseudonym Lewis Herber in *Our Synthetic Environment.* According to Bookchin, the agricultural mass use of chemicals was alienating humans from the natural world. Once again, the intensive farm featured prominently as a dangerous site of environmental and nutritional contamination.[32] Similar opinions were voiced in the British media. Commenting on the 1962 decision by the Court of Appeal to deny intensive broiler houses the status of agricultural buildings, the *Daily Mail* noted: "Now we know what the broiler chicken really is—not a creature of the farm but a product of the factory."[33]

Popular concerns about the 'unnatural' health hazards of intensive food production fused with rising uneasiness about new systems' effects on animals themselves. While most agricultural commentators supported intensive production,[34] some veterinarians and farmers had since the 1950s warned about the detrimental effects of 'artificial' practices and treating animals like 'machines.' However, even sceptical producers rarely called for bans or regulation. Instead, they relied on a theory of natural self-regulation, which stated that pushing animals too hard would result in lower productivity, economic losses, and production adjustments. Described by Abigail Woods, this equation of animal productivity—thrift—with adequate welfare failed to satisfy a growing number of non-agricultural observers.[35]

The introduction of continental veal production systems proved particularly contentious. In 1960, major British newspapers reported that the new intensive systems produced 'white' meat by inducing anaemia in young calves with diets deficient in iron, by rearing animals in total darkness in small wood crates, and by bleeding them.[36] Resulting outrage sparked an inconclusive official enquiry by the Parliamentary Animal

[32] Lewis [Pseudonym for Murray Bookchin] Herber, *Our Synthetic Environment* (New York: Knopf, 1962).

[33] "The Farm Factory", *Daily Mail*, 04.07.1962, 1; see also: Clifford Selly, "Chicken Farm Or Factory?", *Observer*, 08.03.1959, 3.

[34] See, for example, parallel defences of intensive production in the *Daily Mail*, Alan Exley, "Twelve Week Wonder", *Daily Mail*, 15.04.1959, 11–12; A.G. Street , "How I hate the chicken", *Daily* Mail, 09.04.1960, 11; Peter Black, "Conversation with a broiler." *Daily Mail*, 16.08.1963, 6; "On a Yorkshire rabbit farm. A parallel to the broiler industry", *Guardian*, 25.11.1960, 18.

[35] Abigail Woods, "From cruelty to welfare," 17.

[36] "Calves 'Reared in Broiler Houses'", *Guardian*, 16.06.1960, 1; "Calves' Growth Encouraged By Draining Blood", *Guardian*, 30.04.1960, 2; "Comment: The Fatted Calves", *Daily Mail*, 16.07.1960, 1.

Welfare Group, which criticised practices as inhumane but could not legally fault them.[37] The group's findings reinforced concerns about existing welfare legislation: relying on self-regulation via thrift and the prosecution of individual cases of excessive cruelty was not enough if economically viable systems like intensive veal production were inherently inhumane. Outrage over veal production also focused public attention on other intensive practices like battery systems for hens and the long-distance transport of live animals. Between December 1962 and January 1963, three consecutive issues of the *Daily Mirror* attacked stalling legislative reform regarding the long-distance trade of male bobby calves from the West Country and Scotland for fattening or slaughter in the South of England. According to the *Mirror* and the RSPCA, transported calves were never fed, suffered multiple instances of physical abuse, and frequently died before reaching abattoirs or markets.[38]

Reacting to rising concerns about so-called broiler systems, Labour MP John Dugdale, former private secretary to Clement Attlee and relative of ex-Minister of Agriculture Thomas Dugdale, introduced the 1960 Animals (control of intensified methods of food production) Bill. Dugdale's private bill attempted to ensure humane living conditions by regulating the construction of new buildings to give animals more space.[39] The Bill was supported by the inter-party Parliamentary Animal Welfare Group.[40] However, without government support, it failed to secure a second hearing. The Ministry of Agriculture Fisheries and Food (MAFF) objected to the Dugdale Bill because of the "extreme difficulty of trying to regulate the detailed conditions under which farm animals are to be kept."[41] In a sign of how polarised welfare debates were becoming, MAFF officials successfully lobbied against a more general welfare enquiry "because of the unlikelihood that generally acceptable conclusions would ever be reached on a subject which generates so much emotion."[42]

[37] "Calf-rearing for veal", *Guardian*, 21.07.1960, 5.

[38] "This sad, sad business', *Daily Mirror*, 31.12.1962, 10–11; "The Evidence about this sad, sad business", *Daily Mirror*, 01.01.1963, 10–11; "This sad, sad business", *Daily Mirror*, 02.01.1962, 10–11.

[39] "Animals (Control of Intensified Methods of Food Production)," *Hansard* Vol. 630 (23.11.1960); Woods, "Cruelty to Welfare," 16.

[40] "Animals (Control of Intensified Methods of Food Production)," *Hansard* Vol. 630 (23.11.1960).

[41] TNA MAF 293/169 Minute: Mr Hutchison to ES Virgo (05.03.1964).

[42] TNA MAF 293/169 Minute: Mr Hutchison to ES Virgo (05.03.1964).

The 'emotionality' of welfare concerns was exacerbated by intensive methods' alleged foreignness.[43] Fusing wartime tropes of British compassion with *Kulturkritik* of alienating technology, contemporaries highlighted the foreign origins of intensive 'American' broiler systems, 'Dutch' white veal, and 'German' pig sweat boxes.[44] Intensive systems were presented as a threat to 'core' British values like 'freedom' and 'decency.' In 1960, the Archbishop of Canterbury considered the increasing adoption of Dutch intensive calf production an "absolute horror" and a "blow to decent feeling."[45] Commentators in the *Daily Mail* joined him in calling the alien systems unchristian:

> No one can doubt that, given the choice, the calves would opt for the green fields, the sunlight, and the winds of heaven, even though the meadow was sparse and the weather unfriendly. They love their freedom too. (…). The real indictment of the 'broiler' system is that it is so unnatural that it outrages ordinary human feelings. (…) there is a moral gulf between mechanising a plough and mechanising an animal. (…) we shall all lose something of mind and spirit if we begin to abandon immemorial pastoral ways and scenes merely, for example, to please the consumers of veal and ham pies.[46]

In Parliament, future Labour Minister of Agriculture, Thomas Frederick Peart, reacted to the failure of the 1960 Dugdale bill by stressing the 'British' values of kindness and tolerance: "there might be some people who thought more of animals than of human beings but British people were kind and tolerant. He had always found when abroad that standards of conduct were not comparable with those existing in Britain in animal welfare."[47] According to the *Daily Mail*, 90 per cent of readers opposed veal calf production because they defined "'cruelty' as something more

[43] Links between welfare and nationalist discourse remain strong; see ongoing work on Brexit and animal welfare campaigning by Reuben Message.

[44] "Calves' Growth Encouraged By Draining Blood", *Guardian*, 30.04.1960, 2; Elspeth Huxley, *Brave New Victuals. Are We All Being Slowly Poisoned? A Terrifying Enquiry into the Techniques of Modern Food Production* (London: Panther Books [1965] 1967), 27–29; RSPCA Archives, CM/57 RSPCA Council Minutes, Meeting of the Council, 25.07.1968, 2.

[45] "Comment: The Fatted Calves", *Daily Mail*, 16.07.1960, 1.

[46] "Comment: The Fatted Calves", *Daily Mail*, 16.07.1960, 1.

[47] "Calves' growth encouraged 'by draining blood'", *Guardian*, 30.04.1960, 2; see also: "Calves 'Reared on Broiler System'", *Daily Mail*, 17.05.1960, 11.

than pain, starvation or persistent bad treatment. The 'cruelty' to these calves lies in the deprivation of their natural freedom." [48]

The fact that intensive production took place in 'the countryside' as a highly stylised seat of upper-middle class English identity further heightened the perceived moral and physical threat of 'alien' factory farms.[49] Coinciding with a boom of countryside writing and motorised tourism, protests against factory systems blurred with protests against urban encroachment and attempts to preserve rural heritage in the form of hedgerows, country houses, and traditional ways of life.[50] In 1962, *Punch* satirised the visit of city girl Linda from Walham Green to Uncle Henry at Jollity Farm, a "burly man with a bundle of £5 notes thrust carelessly into his hatband," whose farm specialises in destroying British countryside hedgerows and killing woodpigeons:

> By the end of three days Linda suspected that her uncle was a god (…). 'Uncle Henry', she suggested as she trotted behind him to check the day's takings, 'why don't you make square hens? Then you could pack four to a cage.' And how proud she was when Uncle Henry patted her head and said bless his buttons but he might have a try yet.[51]

Looking at the US, the *Daily Mail* warned that the "transition from farm to factory, and the substitution of the natural by the artificial" would lead to the "shriek of the factory whistle" replacing the traditional "call of the land."[52]

Critics' portrayal of intensive animal protection as a systemic threat to health, the environment, animal welfare, British values, and 'the

[48] "Editorial: The Calves Again", *Daily Mail*, 23.07.1960, 1.

[49] David Lowenthal, "British National Identity and the English landscape," *Rural History* 2/2 (1991), 205–230.

[50] Sophia Davis, "Secluded Suffolk: Countryside Writing, c. 1930–1960," *Island Thinking* (2019), 31–71; Malcolm Chase, "This is no claptrap: this is our heritage," in Christopher Shaw and Malcolm Chase (eds.), *The Imagined Past: History and Nostalgia* (Manchester: Manchester University Press, 1989), 128–146; Catherine Brace, "Looking back: the Cotswolds and English national identity, c. 1890–1950," *Journal of Historical Geography* 25/4 (1999), 502–516; Veldman, *Fantasy*, 215–219; Sean Nixon, "Trouble at the National Trust: Post-war Recreation, the Benson Report and the Rebuilding of a Conservation Organization in the 1960s," *Twentieth Century British History* 26/4 (2015), 529–550; Cassidy, *Vermin*, 166–167; Sayer, "Animal Machines," 488, 495–496.

[51] Angela Milne, "A Plague of Pigmy Shrews", *Punch*, 30.05.1962, 820–821.

[52] "The Farm Factory", *Daily Mail*, 04.07.1962, 1.

countryside' caused increasing tensions with producer interests. Faced with concerns about dewy-eyed calves, the British National Farmers' Union (NFU) responded to cruelty allegations with articles such as "Broiler Veal Not Cruel—Says NFU,"[53] "Calves don't suffer—Mr. Hare,"[54] and "Cruel to their Kind?"[55] Agricultural commentators stressed that poorly treated animals would not be profitable and attempted to defend intensive systems with expert studies and references to allegedly high production standards.[56] In comparison to the "pot-bellied" pre-war animals "with staring coats, housed in filthy hovels,"[57] intensive systems offered a much better life. According to *Farmers Weekly* columnist A.G. Street:

> The charge that such intensive methods of fattening are cruel springs from the inevitable difference between the definition of 'cruelty' according to whether one is country or town bred. To the former—and especially the farmer—cruelty is ill-treatment, especially to the extent that the health and thriving of the animal is adversely affected. But the latter usually add to this what they call lack of consideration. Rightly or wrongly—I think wrongly—townsfolk are apt to invest dumb animals with human minds and hopes and emotions.[58]

Farming organisations were not the only ones to experience pressure in the face of new welfare demands. Rising criticism of intensive farming also challenged traditional animal protection bodies like the RSPCA. During the 1950s, the Society had focused on preventing individual instances of cruelty like the exploitation of circus animals or the so-called ritual slaughter of unstunned animals. More systematic considerations of animal welfare in new agricultural production systems had not been a prominent campaigning focus.[59] Starting in the 1960s, rising grassroots opposition to hunting and agricultural intensification would lead to an adaptation of campaigning goals, criticism of the Society's elite leadership, and a growing fragmentation of ties between the RSPCA and Whitehall. Studying RSPCA Council minutes reveals the growing intensity of internal and external struggles.

[53] "Broiler Veal Not Cruel—says NFU", *FW*, 22.07.1960, 38.

[54] "Calves don't suffer—Mr Hare", *FW*, 29.07.1960, 40.

[55] A.G. Street, "Cruel to their Kind?", *FW*, 30.09.1960, 83.

[56] J. Sandison, "What Suits Calves…", *FW*, 29.07.1960, 62–63; "Broiler Veal Not Cruel—says NFU", *FW*, 22.07.1960, 38

[57] A.G. Street, "Cruel to their Kind?", *FW*, 30.09.1960, 83.

[58] A.G. Street, "Cruel to their Kind?", *FW*, 30.09.1960, 83.

[59] RSPCA Archives, CM/ 50–54, Committee meetings 1954–1960; campaigning was mostly limited to pamphlets and newspaper articles; "Battery Eggs: RSPCA Fights Hen-Cruelty", *Daily Mail*, 16.07.1953, 1.

The issue of so-called field sports (hunting for pleasure) proved particularly contentious. As described by Angela Cassidy, protests against hunting charismatic animals have a long history in Britain. The early twentieth century had seen an increase of popular opposition to 'cruel' forms of hunting. Protests had significant class dimensions. While middle- and lower-class hunting of badgers and otters steadily decreased, fox hunting with hounds became increasingly popular among Britain's upper class. Opposing elite interests posed challenges for establishment charities like the RSPCA. In 1924, internal RSPCA tensions about upper-class hunting practices led to a schism. Disagreeing with the Society's decision not to publicly oppose 'field sports,' disgruntled members formed the League Against Cruel Sports (LACS). Alarmed by LACS activities and growing opposition among RSPCA members, hunting and anti-poaching societies formed the British Field Sports Society (BFSS) as a counter lobby in 1932.[60]

After dying down during the war, conflicts over 'field sports' resurfaced after 1945. Although an attempt to outlaw fox hunting with hounds failed in 1949, the Scott Henderson report on Cruelty to Wild Animals (1951) resulted in a ban of mechanical gin traps in 1954. In the same year, the Protection of Birds Act provided an important legislative template "against any person who 'kills, injures or takes' wild animals."[61] Invigorated by growing public support for conservationist and environmentalist causes, LACS activists disrupted the RSPCA's 1961 annual general meeting with calls for more decisive action against 'blood sports' like foxhunting with hounds, before being removed by the police. The RSPCA's leadership reacted by rapidly expelling protesters from the Society because of their "highly undignified conduct" and instigation of "rowdyism."[62] However, underlying problems refused to disappear.[63]

In addition to member protests against 'field sports,' RSPCA leaders also had to find a way of addressing growing concerns about intensive farming. The RSPCA had already warned in 1959 about the import of

[60] Cassidy, *Vermin*, 164–165; in many ways, LACS continued earlier protest by Henry Salt's Humanitarian League, Tichelar, *Blood Sports*, Chapter 4; Callum C. McKenzle, "The Origins of the British field sports society," *The International Journal of the History of Sport* 13/2 (1996), 177–191.

[61] Quoted according to Cassidy, *Vermin*, 167.

[62] RSPCA Archives, CM/54, Meeting of the Council, 20.07.1961, 1; see also: "New Drive against fox hunting", *Guardian*, 18.06.1961, 11.

[63] "Expulsions from the RSPCA", *Guardian*, 07.08.1961, 9; "Row over foxhunting at RSPCA meeting", *Guardian*, 15.06.1962, 12.

Dutch veal systems to Britain.[64] In 1960, campaigners raised the issue of calves in "broiler houses" at the Society's annual general meeting and demanded that the RSPCA "'strangle' the practice in its infancy."[65] Bowing to popular pressure, the RSPCA sent a letter to MAFF "urging the discontinuance of broiler and battery houses, and asking that so long as these systems persist, all battery produced eggs should be appropriately stamped" as well as fowl meat produced in "broiler plant' systems."[66] In the same year, the RSPCA Council supported John Dugdale's Animal Bill and subsequent efforts to regulate animal transits.[67] Senior Council members like RSPCA vice-chairman Dr Robert Rattray also published pamphlets and provided information for media articles expressing concern about intensive systems and animal health.[68]

However, the Society's leadership was hesitant to depart from established protest practices—even though these proved inadequate to address the new welfare challenges posed by intensive farms. In 1960, RSPCA veterinary inspections had revealed no offences in intensive facilities, which would have been prosecutable with the 1911 Act.[69] The Council also felt that it was impossible to use existing legal instruments to oppose the spread of new intensive calf systems.[70] Rather than jeopardise ties to MAFF with vocal public protest, the Society relied on more respectable forms of lobbying like parliamentary questions and the high-society connections of its elite leadership to push for legal reform. In public, early RSPCA opposition to intensive farming was couched mostly in moral rather than in ethological, health, or environmental terms. In 1961, RSPCA pamphlets attacked the "intensive exploitation of food animals": "It is the moral wrong in this practice to which we object. It is wrong to treat living things solely with a view to making money out of them and without reasonable

[64] RSPCA Archives, IF/25/1 RSPCA Intensive Farming 2 of 2, RSPCA Annual Report for 1960, 218.

[65] "Calves 'reared in broiler houses'", *Guardian*, 16.06.1960, 1.

[66] RSPCA Archives, IF/25/1 RSPCA Intensive Farming 2 of 2, RSPCA Annual Report for 1960, 221.

[67] "Animals (Control of Intensified Methods of Food Production)," *Hansard* Vol. 630, 23.11.1960; RSPCA Archives, CM/55, Meeting of the Council, 21.02.1963, 3.

[68] "Battery Eggs: RSPCA Fights Hen-Cruelty", *Daily Mail*, 16 July 1953, [1] See also: RSPCA Archives, IF/25/1 RSPCA Intensive Farming 2 of 2, RSPCA Annual Report for 1961, 231.

[69] "Calves 'reared in broiler houses'", *Guardian*, 16.06.1960, 1.

[70] RSPCA Archives, IF/25/1 RSPCA Intensive Farming 2 of 2, Annual Report for 1960, 220.

consideration for them."[71] Dissatisfied with what they interpreted as overly cautious campaigning, more radical RSPCA members began to call for vocal wholesale opposition to agricultural intensification.[72]

Intensive animal production systems had thus already become focal points of public protest by the time Ruth Harrison started writing *Animal Machines*: while some critics stressed systems' alleged dangers for human, environment, and animal health, others stressed their 'alien' origins and detrimental effects on animal welfare, British values, and 'the countryside.' It was also becoming clear that existing case-by-case legal mechanisms for animal protection were no longer adequate to deal with the systemic welfare issues raised by intensive animal production. Dissatisfied with government politics and cautious campaigning by organisations like the RSPCA, a growing number of activists began to demand a radical overhaul of welfare politics. What was needed was an event to tie together the disparate strands of contemporary protest movements and galvanise sufficient societal pressure for reform. The runaway success of *Animal Machines* in March 1964 achieved just that.

[71] RSPCA Archives, IF/25/1 RSPCA Intensive Farming 2 of 2, RSPCA Annual Report for 1961, 231; see also: RSPCA Archives, IF/25/1 RSPCA Intensive Farming 2 of 2, Part 2, Extracts from RSPCA Minutes relating to intensive farming, 13.03.1963.

[72] RSPCA Archives, IF/25/1 RSPCA Intensive Farming 2 of 2, Part 2, Extracts from RSPCA Minutes relating to intensive farming, 19.01.1961, item 8.

CHAPTER 6

Staging Welfare: Writing *Animal Machines*

Animal Machines had been crafted to produce maximum impact. Ruth Harrison began to systematically collect material for her book in 1961. Over the next three years, she read scientific publications on animal behaviour, visited British farms, and corresponded with manufacturers, parliamentarians, and other campaigners—the most prominent of whom was Rachel Carson.[1] Hardly any of her findings were novel. As the past two chapters have shown, numerous activists, scientists, and media outlets had already criticised issues ranging from behavioural constraints on farms to chemical residues in animal products. Rather than providing shocking revelations, *Animal Machines*' impact was instead based on its RADA-trained author's ability to stage and fuse existing concerns. The result was an easy-to-read, compelling moral narrative that focused readers' attention on highly effective examples of alleged cruelty, introduced scientific concepts about animal welfare, and contrasted a dystopian 'factory farm' with a romanticised countryside.

Accessing the latest ethological literature and contacting relevant experts was not straightforward for someone without access to agricultural, policy, or academic networks. On farms, Harrison noted that "farmers were

[1] FACT Files, DB, Box, Material for 'Animal Machines'.

C. Kirchhelle, *Bearing Witness*, Palgrave Studies in the History of Social Movements,
https://doi.org/10.1007/978-3-030-62792-8_6

astonishingly unaware that their methods were questionable."[2] Meanwhile, government officials either assured her that everything was fine or claimed that the 1911 Protection of Animals Act permitted no further regulatory action to enhance animal welfare, beyond preventing immediate physical harm.[3]

Other responses to Ruth Harrison's information requests were more productive and provided her with promising leads for crafting her narrative. One of these early leads was the humane stunning and killing of animals. Harrison had been shocked to learn that the slaughtering of animals in large abattoirs was not necessarily painless. Following correspondence with the Council of Justice to Animals and Humane Slaughter Association, she received information on electric stunners. Although stunning was by no means universal in secular abattoirs, Harrison followed contemporary campaigns by the Council of Justice and the RSPCA by targeting the unstunned "Jewish and Mohammedan Slaughter of Food Animals" in early drafts of her book.[4]

White veal production (see Chap. 5) quickly became a second pillar of her book's attack on 'factory farms.' Harrison's moral outrage at the intensive production of anaemic calves in what would become Chapter Five of *Animal Machines* is palpable.[5] Production practices had already featured prominently in the Crusade Against All Cruelty to Animals' letterbox leaflet to Harrison. In his testimony for the Crusade, "Suffolk farmer" and *Daily Mail* contributor Laurence Easterbrook highlighted the "wretched trade['s]" unBritish characteristics—modern systems of veal and broiler husbandry "might well have been devised by Hitler."[6] Harrison decided to conduct further research and requested information from Gwendolen 'Gwen' Barter, who had come to national fame after disrupting the RSPCA's 1961 annual general meeting to protest against

[2] Colin Spencer and Spike Gerrel, "A rare breed at the factory farm", *Guardian*, 03.11.1990, A19.

[3] FACT Files, DB, Box, Material for 'Animal Machines', Chapter Eight. Quantity versus Quality, 122, JH Tucker to Mrs Harrison (20.08.1962); Chapter Five. Veal Calves. Copy WMF Vane to Sir Henry Studholme (April 1961).

[4] FACT Files, DB, Box, Material for 'Animal Machines', Chapter Five. Veal Calves. Council of Justice to Animals and Humane Slaughter Association, 'Jewish and Mohammedan Slaughter of Food Animals'; Council of Justice to Animals and Humane Slaughter Association to Ruth Harrison (29.01.1962).

[5] Ruth Harrison, *Animal Machines—New Edition* (Wallingford and Boston: CABI, 2013), 85–105.

[6] FACT Files, DB, Box, Material for 'Animal Machines', Chapter Five. Veal Calves, Laurence Easterbrook, 'Stop This Wretched Trade', 1–2.

'field sports.' Barter had actively lobbied both the RSPCA and MAFF to stamp out veal production and passed on relevant correspondence to Harrison,[7] who supplemented MAFF statements with published industry, veterinary, and personal descriptions of production practices, as well as with images taken during farm visits.[8]

This combination of written and visual depictions of intensive production proved extremely effective. Similar to Greenpeace 'mindbombs,' which eschewed complex content in favour of simple visual messaging like the clubbing of a baby seal,[9] Harrison made extensive use of images for her attack on 'factory farms.' The cover image of the 1964 edition was a linocut depicting a spinning cogwheel whose centre consisted of a cow's head and whose individual cogs consisted of alarmed chickens' heads (Image 6.1). Upon opening the book, readers were greeted by a cartoon of three farmers trying to convince a herd of dairy cows, who had clearly never ventured outside, to eat grass: "It's GRASS y'fools—you're supposed to EAT it—Remember?"[10]

The book's core message of unnatural, unhealthy, and cruel intensive farming was condensed in a "Pictorial Summary."[11] The summary was strategically placed in the middle of *Animal Machines* where it functioned as a narrative hinge. The hinge connected descriptions of intensive production systems for poultry, calves, and other species in the first half of the book with the second half's analysis of food quality and animal legislation and resulting call for welfare reform. Opening with *Animal Machines*' most striking image—a picture of a wide-eyed calf staring out of a small, dark crate[12]—the pictorial summary's 24 high-quality photographs contrasted intensive and alternative methods. Staged contrasts were not subtle: an image of a dog overlooking a bucolic countryside farm with meadows full of sheep was contrasted with a concentration camp-like image of utilitarian broiler bar-

[7]FACT Files, DB, Box, Material for 'Animal Machines', Chapter Five. Veal Calves, Gwendolen Barter to Minister of Agriculture (11.08.1960); "Gallant Gal Fails to Free Boxed Fox", *Reading Eagle*, 08.03.1966, 16; a clip with a scene of the protest and an interview with Barter for the ITV Late Evening News on 14.06.1961 can be seen at "Foxhunting Protest at Rspca Meeting", *ITN Source. JISC MediaHub.*

[8]Ruth Harrison, *Animal Machines—New Edition*, 85–105, 115, 126–133.

[9]Peter Dauvergne and Kate J. Neville, "Mindbombs of right and wrong: cycles of contention in the activist campaign to stop Canada's seal hunt," *Environmental Politics* 20/2 (2011), 192–209.

[10]Ruth Harrison, *Animal Machines—New Edition*, 34.

[11]Ruth Harrison, *Animal Machines—New Edition*, 114–138.

[12]Ruth Harrison, *Animal Machines—New Edition*, 115.

Image 6.1 Cover of *Animal Machines* (Ruth Harrison's personal copy) (image courtesy of Ruth Layton)

racks; a farmer feeding an outdoor flock was contrasted with a white-coated worker in a battery house; a cow nursing her calf in a field was contrasted with shackled calves standing on concrete slats with swollen knee joints. The photographs had either been shot by Dex Harrison or been purchased from farming magazines and other campaigning organisations. Each picture was accompanied by a brief take-home message. Some of these were pithy one-liners as with a photograph of the interior of a pig sweathouse, which was simply titled "Phew-w...w!"[13] Others contained more detailed reflections on the design of production and slaughter facilities as well as on their welfare and environmental impacts (Image 6.2). The pictorial summary drove

[13] Ruth Harrison, *Animal Machines—New Edition*, 138.

Image 6.2 *Animal Machines'* image of a veal calf looking out of its crate with the neighbouring crate shut (image courtesy of Jonathan Harrison)

its message home with three questions: "in degrading these animals are we not in fact degrading ourselves? (…) At what point do we acknowledge cruelty? (…) Can these unhealthy animals possibly make healthy human food?"[14]

In addition to its effective use of visual imagery to summarise its core messages, what also set Harrison's book apart from other attacks on 'factory farming' was its popularisation of ethological research on animal instincts and the frustration of instincts on intensive operations. For her chapter on "Cruelty and Legislation," Harrison relied heavily on Universities Federation for Animal Welfare (UFAW) publications. Although she did

[14] Ruth Harrison, *Animal Machines—New Edition*, 114.

not reference newer work on stress or the humane treatment of laboratory animals (see Chap. 4), Harrison was very interested in a 1948 publication by Oxford zoologist John R. Baker. Addressing the "scientific basis of kindness to animals," Baker claimed that "it is probable that there is some degree of correlation between intelligence and capacity to suffer."[15]

For Harrison, the question of outwardly healthy animals' capacity to suffer in intensive production systems was of central importance to refuting the equation of animal productivity—thrift—with welfare.[16] It was also key to her argument for legislative reform by showing that the cruelty definitions set out in the 1911 Protection of Animals Act were inadequate. However, her resulting interest in issues like boredom, 'abnormal' behaviour, and animal 'vices' like feather pecking or tail biting was not shared by leading continental ethologists (Chap. 4). Replying to a 1961 letter from Harrison, Konrad Lorenz noted: "we need not torment our conscience too much about unavoidable cruelty though of course we are in honour bound to avoid all avoidable cruelty."[17] Lorenz did not think that "the heavy domesticated breed of chicken suffer seriously either under the measures taken to produce quick growth or under the conditions of hens kept in batteries."[18] Chickens did "not 'understand' the situation when their fellows are being slaughtered or lying dead."[19] Willing to "incur the danger of your thinking me callous and cruel," Lorenz asserted that maltreated animals would not thrive: "The claim that birds which are distinctly unhappy would not lay many eggs is, in my opinion, perfectly justified."[20]

Rather than deter her, Lorenz's reply intensified Harrison's efforts to find sympathetic research. For her book, Harrison drew on UFAW work on instincts, Cambridge veterinary researcher David Sainsbury's research on animal vice, and US naturalist Roy Bedichek's descriptions of nervous and

[15] FACT Files, DB; Box, Material for 'Animal Machines', Chapter Nine: Cruelty + Legislation, 142, Dr John Baker, "The Scientific Basis of Kindness to Animals". Published by UFAW—First Issued June 1948; Reprinted 1951 and 1955, 8.

[16] Harrison was particularly interested in intensive systems' effect on "vice" in animals; FACT Files, DB; Box, Material for 'Animal Machines', Chapter Eight: Quality versus Quality, 116, Pamphlet.

[17] FACT Files, DB; Box, Material for 'Animal Machines', Chapter Three: Poultry Packing Stations, Konrad Lorenz to Mrs Harrison (19.10.1961).

[18] FACT Files, DB; Box, Material for 'Animal Machines', Chapter Three: Poultry Packing Stations, Konrad Lorenz to Mrs Harrison (19.10.1961).

[19] FACT Files, DB; Box, Material for 'Animal Machines', Chapter Three: Poultry Packing Stations, Konrad Lorenz to Mrs Harrison (19.10.1961).

[20] FACT Files, DB; Box, Material for 'Animal Machines', Chapter Three: Poultry Packing Stations, Konrad Lorenz to Mrs Harrison (19.10.1961); Lorenz, however, opposed veal crates.

bored battery hens.[21] She also began contacting other British ethologists like fellow Quaker William Homan Thorpe, whose acknowledgement of animal cognition and welfare views mirrored her own (Chap. 4). While it is unclear whether resulting contacts were established in time to influence the writing of *Animal Machines*,[22] they would later help Harrison secure positions on government committees, scientific backing for her campaigning, and high-profile members for her charity (see Parts III and IV).

Now often forgotten, a third major theme of *Animal Machines* centred on intensive agriculture's environmental and health impacts. The publication of *Silent Spring* (1962) in the US one year into her research had a significant effect on Harrison's own writing process. Carefully studying public reactions to *Silent Spring*[23] and citing the book in her chapter on "Quantity versus Quality,"[24] Harrison was keenly aware of *Silent Spring*'s impact on British attitudes towards agricultural chemicals. The similarities between *Silent Spring* and her own book project must have been obvious, and Harrison established contact with Carson via the editorial office of the *New Yorker* on November 9, 1962.[25] In her letter to Carson, Harrison wrote that she was working on a book on "factory farming" and asked for "assistance ... on the nutritional side."[26] Indicating the breadth of her interests, Harrison felt "whereas the cruelty angle causes only momentary interest, the questionable food value of foods so produced might raise a more lasting doubt in people's minds."[27] Describing the rearing of veal

[21] Ruth Harrison, *Animal Machines—New Edition*, 153, 175–184.

[22] Cambridge's Thorpe Papers contain no correspondence between Harrison and Thorpe, and there are also no letters from Harrison in Rice University's Huxley Papers. However, Thorpe's address book from this time lists Harrison's home address; Cambridge University Library, William H. Thorpe Papers GBR/0012/MS Add.8784, box 3, black address book; the other 'founder' of ethology, Nikolaas Tinbergen, remained sceptical of animal welfare activists like Harrison; correspondence with Marian Dawkins (07.08.2015).

[23] FACT Files, DB; Box, Material for 'Animal Machines', Chapter Eight: Quality versus Quantity, 134, Cutting: Minister Deplores Alarm Over Farm Chemicals (21.03.1963).

[24] FACT Files, DB; Box, Material for 'Animal Machines', Chapter Eight: Quality versus Quantity, Note iii.

[25] The UK version of *Silent Spring* only appeared in 1963; Yale Beinecke Library [in the following YBL], Rachel Carson Papers [in the following RCP], YCAL, MSS 46, Series II, General Correspondence [in the following GC], Box 103, Folder 1952, Ruth Harrison to Rachel Carson c/o The New Yorker (09.11.1962).

[26] YBL, RCP, YCAL, MSS 46, Series II, GC, Box 103, Folder 1952, Harrison to Carson c/o the New Yorker (09.11.1962), 1.

[27] YBL, RCP, YCAL, MSS 46, Series II, GC, Box 103, Folder 1952, Harrison to Carson c/o the New Yorker (09.11.1962), 1.

calves, Harrison asked Carson "whether animals reared in an unhealthy way can possibly produce healthy food"[28] and whether eating animals that had come into contact with antibiotics or hormones was safe.

Responding on November 23, Rachel Carson admitted being "quite appalled by your letters describing matters that I had known very little about."[29] Although Carson was unsure whether unhealthy animals produced unhealthy meat, she knew "that many Doctors [sic] feel that the fact that we all get small doses of antibiotics from eggs, meat, and so on has something to do with the fact that so many bacteria have become resistant to these drugs."[30] In the case of insecticides, Carson also noted that a "tolerance has been set for the occurrence of insecticide residues in meat."[31] It was her belief that "this situation should be changed,"[32] and she referred Harrison to British physician Franklin Bicknell's 1960 publication *Chemicals in Your Food.*[33]

Ruth Harrison must have found this rapid response encouraging because she sent six draft chapters to Carson in May 1963 and asked whether Carson would like to contribute a preface or foreword to *Animal Machines.* Well aware that intensive production systems were only beginning to establish themselves in the UK, Harrison hoped that Carson would help her prevent their further spread:

> I realise that you have not personally studied this subject, but I hope that these chapters will show you that as each set of conditions becomes established it becomes more and more difficult to change them. I feel that it is most important to explain to the individual what is happening, and to try to make him aware of his responsibility in allowing it. Both physically and morally that is indeed a poor heritage to pass on to our children.[34]

[28] YBL, RCP, YCAL, MSS 46, Series II, GC, Box 103, Folder 1952, Harrison to Carson c/o the New Yorker (09.11.1962), 2.

[29] YBL, RCP, YCAL, MSS 46, Series II, GC, Box 103, Folder 1952, Rachel Carson to Ruth Harrison (23.11.1962).

[30] YBL, RCP, YCAL, MSS 46, Series II, GC, Box 103, Folder 1952, Carson to Harrison (23.11.1962).

[31] YBL, RCP, YCAL, MSS 46, Series II, GC, Box 103, Folder 1952, Carson to Harrison (23.11.1962).

[32] YBL, RCP, YCAL, MSS 46, Series II, GC, Box 103, Folder 1952, Carson to Harrison (23.11.1962).

[33] Franklin Bicknell, *Chemicals in Food.*

[34] YBL, RCP, YCAL, MSS 46, Series II, GC, Box 103, Folder 1952, Ruth Harrison to Rachel Carson (07.05.1963), 1.

According to Harrison, Sydney Jennings, former President of the British Veterinary Association (BVA), was "backing [her] most staunchly" and had "been kind enough to 'vet' all my husbandry facts."[35]

However, Carson stalled. Referencing her increased workload in the wake of President Kennedy's Science Advisory Committee's report on pesticides, she asked Harrison to send a complete manuscript so that Carson's friend, Christine Gesell Stevens, could check it. Stevens was an important contact. The daughter of University of Michigan physiologist and laboratory animal campaigner Robert Gesell, Stevens was in close contact with both US and British animal protection organisations. Drawing on the resources of her husband, real estate baron Roger Lacy Stevens, she had founded the Animal Welfare Institute (AWI) in 1951. The AWI was based in the Stevens-owned Empire State Building and had just helped push the 1958 US Humane Slaughter Act through Congress.[36] During the 1960s, the AWI lobbied for what became the 1966 US Laboratory Animal Welfare Act, farm animal welfare, and a ban of whale hunting.[37]

Only after Stevens' check was complete did Carson agree to write the foreword in July 1963: "Both Mrs. Stevens and I are much impressed with what you have done, and are delighted that you have undertaken to describe and document this situation for the public."[38] Responding to Carson's suggestion of a further preface from a British expert, Ruth Harrison confirmed that Sydney Jennings had agreed to provide such a preface. *Animal Machines* had also received approval from *Chemicals in Your Food* author Franklin Bicknell. Significantly, Harrison also agreed to make certain revisions in response to Christine Stevens' comments and cut passages dealing with Jewish slaughter practices, a topic which would have alienated some readers and could have detracted from the book's wider arguments:

[35] YBL, RCP, YCAL, MSS 46, Series II, GC, Box 103, Folder 1952, Harrison to Carson (07.05.1963), 1.

[36] YBL, RCP, YCAL, MSS 46, Series II, GC, Box 103, Folder 1952, Rachel Carson to Ruth Harrison (10.06.1963); Carson herself served as an adviser to Stevens' Animal Welfare Institute; Wolfgang Saxon, "Christine Stevens, 84, a Friend to the Animals", *New York Times*, 15.10.2002, 25; Robert Kirk, "Science and humanity".

[37] Adam Bernstein, "Christine Stevens Dies", *Washington Post*, 11.10.2002.

[38] YBL, RCP, YCAL, MSS 46, Series II, GC, Box 103, Folder 1952, Rachel Carson to Ruth Harrison (01.07.1963).

> My only hostility is to the broiler industry in the United States I will prune to make this clear. I agree with her comment on the possible detraction from the main theme by the mention of Kosher slaughter. I will omit this. I would disagree with her acceptance of Kosher slaughter as being in any way humane, but that is beside the point.[39]

Indirectly, Harrison also asked for Carson's help in publishing *Animal Machines* in the US. In July 1963, Harrison informed Carson that her British publishers were considering collaborating with US publishers Devin-Adair on an American version of *Animal Machines*.[40] After Devin-Adair rejected the proposal, Houghton Mifflin, the publishers of *Silent Spring*, indicated that they were interested but insisted that Harrison rewrite *Animal Machines* with the help of an American writer. The project of an American *Animal Machines* was later quietly abandoned.[41] Potential hopes that Carson might help with such a project were precluded by her death in April 1964.[42]

Carson's foreword reached Harrison on August 15, 1963, and resulted in profuse thanks from Harrison, who "couldn't have wished for a better Foreword."[43] Carson strategically reinforced *Animal Machines*' dystopian call to action. Drawing on ethical, environmental, and health arguments, she painted a hyperbolic opposition between a romanticised, pre-modern pastoral and modern intensive farming:

> The modern world worships the gods of speed and quantity, and of the quick and easy profit, and out of this idolatry monstrous evils have arisen. Yet the evils go long unrecognized. Even those who create them manage by some devious rationalizing to blind themselves to the harm they have done society. As for the general public, the vast majority rest secure in a childlike faith that 'someone' is looking after things—a faith unbroken until some public-spirited person with patient scholarship and steadfast courage, presents fact that can no longer be ignored.[44]

[39] YBL, RCP, YCAL, MSS 46, Series II, GC, Box 103, Folder 1952, Ruth Harrison to Rachel Carson (10.07.1963).

[40] YBL, RCP, YCAL, MSS 46, Series II, GC, Box 103, Folder 1952, Ruth Harrison to Rachel Carson (10.07.1963).

[41] YBL, RCP, YCAL, MSS 46, Series II, GC, Box 103, Folder 1952, Ruth Harrison to Rachel Carson (14.10.1963).

[42] YBL, RCP, YCAL, MSS 46, Series II, GC, Box 103, Folder 1952, Harrison to Carson (14.10.1963).

[43] YBL, RCP, YCAL, MSS 46, Series II, GC, Box 103, Folder 1952, Ruth Harrison to Rachel Carson (22.08.1963); Rachel Carson to Ruth Harrison (15.08.1963).

[44] YBL, RCP, YCAL, MSS 46, Series I, Writings, Box 95, Folder 1669, Preface by Rachel Carson for Animal Machines by Ruth Harrison, 1; see also: Marc Bekoff and Jan Nystrom, "The Other Side of Silence," 192.

According to Carson, the "pastoral scenes in which animals wandered green fields or flocks of chickens scratched contentedly for their food" had been replaced by "factorylike buildings in which animals live out their wretched existence."[45] As a biologist, Carson found it inconceivable that such animals could produce healthy food. Intensive establishments were regularly swept through with diseases and were "kept going only by the continuous administration of antibiotics."[46] However, health concerns were only one element of the argument against factory farming:

> The final argument against the intensivism now practiced in this branch of agriculture is a humanitarian one. ... It is my belief that man will never be at peace with his own kind until he has recognized the Schweitzerian ethic that embraces decent consideration for all living creatures—a true reverence for life.[47]

Ultimately, Carson hoped that *Animal Machines* would "spark a consumers' revolt of such proportions that this vast new agricultural industry will be forced to mend its ways."[48]

Securing a foreword by Rachel Carson—whose name appeared more prominently on the cover of *Animal Machines* than Harrison's[49]—and a preface by ex-BVA president Sydney Jennings was a major publicity coup for Ruth Harrison. Authored by an unknown layperson, her book was bound to profit from endorsement by well-known experts. Harrison's second major public relations coup came in late 1963, when she agreed to publish two feature articles on 'factory farming' in the *Observer*.[50] Having failed to interest a "top television documentary film make[r]"[51] in her

[45] YBL, RCP, YCAL, MSS 46, Series I, Writings, Box 95, Folder 1669, Preface by Rachel Carson for Animal Machines by Ruth Harrison, 1.

[46] YBL, RCP, YCAL, MSS 46, Series I, Writings, Box 95, Folder 1669, Preface by Rachel Carson for Animal Machines by Ruth Harrison, 2.

[47] YBL, RCP, YCAL, MSS 46, Series I, Writings, Box 95, Folder 1669, Preface by Rachel Carson for Animal Machines by Ruth Harrison, 3.

[48] YBL, RCP, YCAL, MSS 46, Series I, Writings, Box 95, Folder 1669, Preface by Rachel Carson for Animal Machines by Ruth Harrison, 3.

[49] I am indebted to Dmitriy Myelnikov for this observation.

[50] YBL, RCP, YCAL, MSS 46, Series II, GC, Box 103, Folder 1952, Ruth Harrison to Rachel Carson (14.10.1963), backside of letter.

[51] Colin Spencer and Spike Gerrel, "A rare breed at the factory farm", *Guardian*, 03.11.1990, A19; David Attenborough has no knowledge of being contacted by Ruth Harrison, Correspondence with Sir David Attenborough (19.08.2015).

work, the *Observer* articles were an ideal way to promote *Animal Machines* and sensitise the British public to animal welfare issues.

Harrison's *Observer* articles appeared right ahead of the publication of *Animal Machines* in March 1964. Summarising *Animal Machines*' main arguments and titled "Inside the animal factories"[52] and "Fed to Death,"[53] they were widely advertised as "a disturbing survey" of "animals as food machines."[54] The articles themselves included pictures and vivid descriptions of conditions in 'factory farms'[55] and modern abattoirs. Addressing pig, poultry, and veal husbandry, Harrison asked whether the price society was paying for cheap animal products was "not too high."[56] According to Harrison, "the factory farmer and the agri-industrial world behind him [sic]" only acknowledged "cruelty ... where profitability ceases."[57] However, such an equation of animal productivity and welfare was fundamentally flawed in an age of antibiotics, which could keep animals on their feet despite inadequate welfare. Harrison then attacked animals' cramped living conditions on intensive farms, the perpetual twilight in many buildings, the debeaking of poultry, inadequate stunning prior to animals' scalding and slaughter, and official complacency.

In addition to welfare concerns, the articles addressed potential health hazards resulting from 'factory farming.' According to Harrison, it was common for young birds suffering from respiratory diseases or cancer to end up on consumers' tables, with the birds' ill health masked by antibiotics. Unsurprisingly, the fattening of tethered calves in darkened sties with slatted concrete floors provoked Harrison's particular ire. According to Harrison, calves' diets consisted almost "exclusively of barley, with added minerals and vitamins, antibiotics, tranquilisers and hormones."[58] Living in these conditions, some calves became blind, and many suffered from liver damage and pneumonia: "their muscles become flabby and they put on weight rapidly, *but they are not healthy*."[59] Using more and more antibiotics to keep animals alive, farmers and veterinarians were actively contrib-

[52] Ruth Harrison, "Inside the animal factories", *Observer*, 01.03.1964, 21–22.

[53] Ruth Harrison, "Fed to Death", *Observer*, 08.03.1964, 21 and 28.

[54] "Commercial Observer", *Times*, 28.02.1964, 17; "Commercial Observer: 'Animals as Food Machines. A disturbing survey", *Daily Mail*, 28.02.1964, 12.

[55] Harrison, "Inside the animal factories", 21.

[56] Harrison, "Inside the animal factories", 21.

[57] Harrison, "Inside the animal factories", 21.

[58] Harrison, "Fed to Death", 21.

[59] Harrison, "Fed to Death", 21.

uting to a race "between disease and new drugs."[60] The results of this race were antibiotic resistance and residue-laden "tasteless meat"[61]—an admittedly odd comment from someone who had probably never eaten meat.[62]

Harrison's joint staging of environmental, welfare, health, and moral concerns was extremely successful. By addressing the alleged dangers of 'factory–farmed' meat in combination with its welfare implications, she was able to turn 'factory farms' into a focal point of contemporary concerns about technological alienation from nature and resulting effects on the nation's physical and moral health. Whereas *Animal Machines* is now mostly remembered for its welfare message, a closer reading of the book and its origins shows the many intellectual roots connecting it to the wider ferment of post-war environmentalist, conservation, and peace activism. Understanding these roots is important in terms of both *Animal Machines*' wider reform message and its ability to galvanise public protest. Harrison and Carson saw animal welfare and environmental reform as two sides of the same coin. Although *Silent Spring* and *Animal Machines* focused on different core messages, it would be wrong to limit either book to a single message. While DDT overuse was one of several concerns voiced by Carson, Harrison criticised 'factory farms' for both their effects on animal welfare and their wider impact on human health, the environment, and societal morals. By challenging prevalent notions of thrift, Harrison was also able to popularise emerging strands of applied ethology, which acknowledged animals' affective states. Breaking with continental ethologists' refusal to engage in 'anthropomorphic speculation,' British ethologists and younger campaigners would use the political momentum created by *Animal Machines* to call for a new form of animal welfare that encompassed physical and affective states.

[60] Harrison, "Fed to Death", 21.

[61] Harrison, "Fed to Death", 21.

[62] I am indebted to Ashley Maher for this observation.

PART III

Impact (1964–1968)

Appearing on March 9, 1964—one day after the second *Observer* article—[1] *Animal Machines* achieved an impact beyond anything Ruth Harrison could have expected. The book not only mobilised protest against new forms of intensive farming but also opened the door for a sustained public and scientific debate about what good welfare actually was. Convened in response to *Animal Machines*, the so-called Brambell Committee proposed concrete welfare improvements. Its report also contained an important annexe by ethologist William H. Thorpe, which set out five essential freedoms for farm animals. While the Thorpe annexe marked a significant step towards defining basic constituents of welfare, resulting hopes for more stringent legal standards proved premature. Despite sustained pressure for new statutory regulations by a coalition of well-known scientists and campaigners, the Ministry of Agriculture Fisheries and Food (MAFF) and industry representatives successfully pushed for a legal framework based on limited statutory standards and loose voluntary codes. Defining and reviewing these codes would be the task of a new Farm Animal Welfare Advisory Committee (FAWAC). Having come to appreciate the power of closed-door official politics, Harrison was determined to gain entrance to the corporatist world of Whitehall advisory committees. This was no easy task. Despite Harrison's determination, the years after 1964 would see her struggle to overcome

[1] TNA MAF 260/351 Minute: JA Barrah to Mr Hensley (25.02.1964).

sexist stereotypes and use her status as a charismatic outsider to successfully network in activist, scientific, and policy circles. Harrison's nomination to FAWAC in 1967 was a major success and would allow her to maintain influence within the rapidly expanding arena of animal welfare politics.

CHAPTER 7

From Author to Adviser: Ruth Harrison and the *Animal Machines* Moment

Animal Machines was an instant bestseller. Five of the main part's eight chapters dealt with cruelty allegations. Of the remaining three chapters, one dealt with the impact of intensive agriculture on food quality, another covered agriculture's ecological impact, and one chapter assessed the state of existing legislation.[1] According to Harrison, intensive animal husbandry was both ethically corrosive and endangered "the physical well-being of the human race."[2] No longer readily identifiable as agriculture, the factory-like nature of broiler agribusiness was a sign of things to come:

> This then is the broiler industry, vast and struggling, a business rather than an agricultural enterprise as we think of agriculture. And the chickens, after nine to ten weeks in these dim, enclosed houses, reach their required weight of 3 ½ lb and are caught, crated and sent to the 'packing station.'[3]

At the end of their short lives, birds were inserted upside down into conveyor belts, their throats cut, their feathers plucked, and their guts eviscerated. Often birds were not stunned prior to the cutting of their throats.

[1] Harrison, *Animal Machines*. [In the following, I use quotations from the 2013 reprint of *Animal Machines*]; Marian Stamp Dawkins, "Why We Still Need to Read *Animal Machines*," *Animal Machines—New Edition* (Wallingford and Boston: CABI, 2013), 3.

[2] Ruth Harrison, *Animal Machines—New Edition* (Wallingford and Boston: CABI, 2013), 40.

[3] Harrison, *Animal Machines—New Edition*, 55.

C. Kirchhelle, *Bearing Witness*, Palgrave Studies in the History of Social Movements,
https://doi.org/10.1007/978-3-030-62792-8_7

Meanwhile, the high throughput of factory farms and slaughterhouses meant that diseased or residue-laden meat might well be reaching consumers.[4] In the case of battery-farmed eggs, Harrison claimed that more and more birds were being held in confined conditions: "only 20 per cent are now on range, whilst 80 per cent have gone indoors."[5] Produced in dimly lit and cramped conditions that were conducive to cannibalism, the supply of 'battery eggs' already outstripped demand and was of dubious quality.[6] The particularly contentious practices of producing white veal meat and broiler beef were treated in two separate chapters (Chap. 6), followed by a less-detailed coverage of rabbits and pigs.[7]

In her conclusion, Ruth Harrison denied that 'factory farms' and cheap meat were in the public interest. According to Harrison, current agricultural research focused only on intensifying indoor systems and neglected alternatives. Meanwhile, annual subsidies of ca. £340,000,000 for an already oversaturated market clearly showed that further productivity increases made no sense.[8] In the face of global population growth, it was more sensible to help areas of malnutrition 'help themselves' than to increase Western overproduction of animal protein.[9] Government interventions such as subsidies should aim to shift farming's focus from quantity to quality. Citing the 1958 German food law and the 1958 US Delaney Clause on carcinogens, Harrison called on the UK government to protect consumers from dangerous chemical residues. She also urged consumers to use their purchasing power to support good husbandry practices and the production of safe food. Although it did not yet amount to a fully fledged 'positive' vision of welfare, Harrison ended *Animal Machines* by calling for "a new charter for animal welfare."[10] Marking the point at which the book transitioned from using terms like "cruelty" to setting out a vision of humane farming with the term "welfare,"[11] the charter made six demands:

[4] Harrison, *Animal Machines—New Edition*, 59-64.
[5] Harrison, *Animal Machines—New Edition*, 65.
[6] Harrison, *Animal Machines—New Edition*, 65-84.
[7] Harrison, *Animal Machines—New Edition*, 85-109.
[8] Harrison, *Animal Machines—New Edition*, 197.
[9] Harrison, *Animal Machines—New Edition*, 198-200.
[10] Harrison, *Animal Machines—New Edition*, 202.
[11] Harrison, *Animal Machines—New Edition*, 202; this is in contrast to claims that *Animal Machines* did not use the term "welfare", Woods, "Cruelty to Welfare," 17.

1. The abolition of battery cages for laying hens.
2. The abolition of current intensive veal production methods.
3. Legislation banning the rearing of animals on deficiency diets.
4. A ban on permanent tethering.
5. A ban on slats.
6. A ban on keeping animals in dim light or darkness.[12]

Fulfilling these demands would shape the next 36 years of Harrison's campaigning.

Although it is important to remember that 'factory farming' was by no means ubiquitous in Britain in 1964, Harrison's effective staging of a dystopian future of machine-like animals and industrialised suffering in the English countryside created a perfect moral storm.[13] In addition to her carefully staged dystopian imaginary and well-orchestrated promotion campaign (Chap. 6), another reason for Harrison's impact was that she managed to avoid public alignment with either the establishment RSPCA or more radical groups like the Crusade Against All Cruelty to Animals. By presenting herself as an ordinary citizen, who was concerned about the health, environmental, and ethical fallouts of intensive farming, Harrison reached a large group of likeminded mainstream readers.[14] The book did not mention Harrison's vegetarianism, family background, or Quakerism and instead strove to mobilise outrage with frequent references to industry practices designed to fool well-meaning consumers.[15]

The *Observer* articles in particular managed to create a period of sustained national indignation about animal welfare, health hazards, and intensive farming. One week after printing the second article, the left-leaning newspaper had received ca. 320 letters responding to Harrison's allegations.[16] Many readers were outraged: Helen Simpson compared animals' suffering to nineteenth-century child labour,[17] Sheila Mitchell

[12] Harrison, *Animal Machines—New Edition*, 202.

[13] Sayer, "Animal Machines," 482–483; Woods, "Rethinking the History of Modern Agriculture".

[14] Bernard E. Rollin, "Animal Machines—Prophecy and Philosophy," *Animal Machines—New Edition* (Wallingford and Boston: CABI, 2013), 11.

[15] Harrison, *Animal Machines—New Edition*, 80, 139, 159, 163; the book's concept of who was responsible for puchasing choices was highly gendered. "Mothers" and "housewives" were addressed twice as frequently as the "consumer."

[16] "Views on animal factories", *Observer*, 15.03.1964, 30.

[17] Helen M. Simpson, "Views on animal factories: Poles apart", *Observer*, 15.03.1964, 30.

demanded labels for products from intensive farms,[18] and Barbara Willard asked fellow readers to imagine their pets incarcerated in factory farms.[19] The RSPCA's Chief Secretary John Hall also praised Harrison's articles.[20] Other prominent supporters were Canon Rhymes of Southwark Cathedral and the Dean of Llandaff. While the former attacked 'factory farming' and advertised *Animal Machines* in his sermons,[21] the latter publicly compared factory farms to Nazi concentration camps, thereby reviving wartime discourse and 'othering' factory farms and animal cruelty as alien, barbaric, and unBritish (see Chap. 6). In a speech covered by the *Daily Mirror* and *Guardian*, the Dean also warned about residues of antibiotics, hormones, and other drugs in British food.[22] The Dean's radical language triggered further appeals calling for an end of antibiotic abuse on unBritish "farm Belsens."[23] In Parliament, Labour MP Joyce Butler launched a consumer-focused inquiry into agricultural chemicals and residues in food.[24]

There was also criticism of *Animal Machines*. Following the first two *Observer* articles, the NFU issued a press statement:

> The statement in Rachel Carson's foreword 'Gone are the pastoral scenes in which animals wandered through green fields' is so absurd that only ten minutes in the countryside proves it untrue. ... It is in presenting some extreme and isolated examples of factory methods of production as if they were typical of British farming methods as a whole that this book is most misleading.[25]

According to the NFU organ *British Farmer*, the *Observer* had joined the "anti-land lobby" by publishing "two articles on intensive production which [give] a grossly distorted picture of British agriculture."[26] The

[18] Sheila M. Mitchell, "Views on animal factories: Label them", *Observer*, 15.03.1964, 30.

[19] Barbara Willard, "Views on animal factories: Try it on the dog", *Observer*, 15.03.1964, 30.

[20] John Hall, "Views on animal factories: changing the law", *Observer*, 15.03.1964, 30.

[21] "Society 'Needs a Common Morality'", *Times*, 08.06.1964, 6; "Attack on factory farming", *Guardian*, 08.06.1964, 3.

[22] "Cruelty War by Church leader", *Daily Mirror*, 10.08.1964, 3; "'Intensive' farming condemned", *Guardian*, 10.08.1964, 3.

[23] "Get rid of farm Belsen", *Observer*, 24.10.1965, 9.

[24] "Hazard to health in food?", *Guardian*, 28.03.1964, 28.

[25] TNA MAF 293/169 NFU News Cycle 705/64/Press 37, "'Animal Machines': NFU Comment on New Book" [released on 09.03.1964] (06.03.1964).

[26] "Feather Heads", *BF*, 28.03.1964, 1.

Observer was further accused of not printing an NFU counter-statement.[27] In a sharply worded editorial, the agricultural magazine *Farmers Weekly* bemoaned:

> Townspeople ... have been given a horrifying picture of the 'animal factories' ... They are given a chilling picture of broiler house concentration camps and packing station Ausschwitzen [sic], of pig 'sweat-boxes'; of darkened torture-chambers for calves, and of animals going blind in intensive beef lots.[28]

If animals were truly suffering, they would not thrive. The magazine also attempted to sunder Harrison's fusion of chemical and welfare critique:

> Intensive animal production is under attack on humanitarian grounds which are often charged with more emotion than facts about its supposed evils. The use of certain farm chemicals is being questioned on a different plane—chiefly, on sober scientific findings about their persistency ... and possible effects on animals and humans.[29]

Critical voices also surfaced in the national media. Similar to the contemporary discrimination faced by Rachel Carson and other female animal activists,[30] the language chosen to attack *Animal Machines* was designed to discredit its author on the basis of her gender. As discussed above, Harrison had skilfully constructed the image of a caring ordinary citizen to mobilise the public. Critics twisted this image into that of an *over*emotional 'housewife,' whose caring nature led her to anthropomorphise animals and whose claims could be downplayed as uneducated and unscientific. Care for animals was contrasted with 'level-headed' (male) scientific knowledge of animal husbandry and veterinary medicine. Writing to the *Observer*, Harrison's later ally, the Cambridge animal health lecturer David Sainsbury, accused Harrison of presenting a "grossly distorted picture of what is *actually* happening."[31] In the *Daily Mail*, agricultural commentator John Winter attempted to discredit Harrison by presenting her views as

[27] "Feather Heads", *BF*, 28.03.1964, 1.

[28] "Techniques in Question", *FW*, 13.03.1964, 43.

[29] "Techniques in Question", *FW*, 13.03.1964, 43.

[30] Michael B. Smith, "'Silence, Miss Carson!' Science, Gender, and the Reception of "Silent Spring," *Feminist Studies* 27/3 (2001), 733-752; Cassidy, *Vermin*, 182-186, Gaarder, *Women*, 41-60;

[31] David Sainsbury, "Views on animal factories: distorted", *Observer*, 15.03.1964, 30

those of a 'fanciful' female mind. Asking whether farmers were "really as cruel as this housewife says," Winter claimed that "it is time somebody sowed some fresh seeds in the fertile mind of Ruth Harrison."[32] According to Winter, Harrison's three-year research "crusade" against intensive production methods had "taught her little about farming."[33] Dismissing Harrison as a "housewife, mother, and vegetarian," Winter thought it logical that (female) city-dwellers would be "horrified" at conditions in even the best abattoirs:

> Nobody who is unaccustomed to it is happy in a slaughterhouse, and the highspeed, mechanised death chambers of packing stations are repellent. This is largely because of their size, and the number of birds involved. But they are infinitely more humane, and hygienic, than the old primitive slaughter methods.[34]

By intensifying production, British farmers were only reacting to public demand for cheap food, which consumers now saw as their "birthright."[35] Most farmers were "kind and humane in their treatment of stock, without being sentimental."[36]

However, even hostile commentators were soon forced to admit that dismissing Harrison as an overly sentimental housewife would not diffuse public outrage about 'factory farming.' Three days after printing John Winter's attack on Harrison, the *Daily Mail* was flooded with letters protesting the newspaper's perceived support of intensive methods. Although one letter applauded Winter's "level-headed criticism of Ruth Harrison's book,"[37] Norman Barr from Sussex claimed, "I'll never have veal or chicken again if John Winter's article is the best defence against *Animal Machines*."[38] Another letter renewed patriotic analogies with the Second World War and Nazi concentration camps:

[32] John Winter, "Are farmers really as cruel as this housewife says?", *Daily Mail*, 09.03.1964, 8.

[33] Winter, "Are farmers really as cruel as this housewife says?".

[34] Winter, "Are farmers really as cruel as this housewife says?".

[35] Winter, "Are farmers really as cruel as this housewife says?".

[36] Winter, "Are farmers really as cruel as this housewife says?".

[37] "Letters—Man, food and animals", *Daily Mail*, 13.03.1964, 10.

[38] "Letters—Man, food and animals", *Daily Mail*, 13.03.1964, 10.

> According to John Winter those who deplore Belsen and other German concentration camps are expected to condone similar conditions for our farm animals so that we may have cheap food as a birthright—and the farmer mink for his wife.[39]

In an editorial comment, the *Daily Mail* acknowledged that a majority of readers who wrote shared Harrison's criticism of factory farm methods.

Criticism of factory farming also appeared in the agricultural media. Writing to *Farmers Weekly* in 1964, K.M. Ropewind challenged battery farms. Why was an industry suffering from overproduction so intent on producing ever more surpluses?[40] According to Mrs F. Belsham from Kent, Harrison's images of intensive methods were "enough to sicken any person with ordinary humane feelings"[41]:

> it would be a diplomatic move ... to try and stop some of these cruelties instead of forever dismissing as sentimental rubbish any attempt by anyone to show up to an ignorant public some of the methods by which their 'cheap food' is produced.[42]

Whether they supported or disagreed with them, it is fair to say that large parts of the British public were familiar with *Animal Machines*' core messages by the end of March 1964. Within three weeks, Harrison had achieved national fame.

Behind closed doors, the British government had been apprehensive about *Animal Machines* for a while. MAFF officials had received review copies of the book around one week ahead of the *Observer* articles. In their comments, officials warned, "there does not seem to be anything new in this book though the way the material is assembled and the publicity that it will get through the 'Observer' will undoubtedly lead to renewed pressure on us."[43] According to MAFF Deputy Secretary W. C. Tame, "we cannot expect that references to the Protection of Animals Act or to the

[39] "Letters—Man, food and animals", *Daily Mail*, 13.03.1964, 10.
[40] K.M. Petter Ropewind, "Battery Birds", *FW*, 27.03.1964, 41.
[41] F. Belsham, "Factory Farming", *FW*, 27.03.1964, 42.
[42] Belsham, "Factory Farming".
[43] TNA MAF 293/169 Minute W.C. Tame to Mr Hutchison (28.02.1964).

theory that animals do not thrive unless they are happy, will be readily accepted"[44]:

> I wonder whether the time has come when we ought to consider seriously—in spite of the obvious difficulties—the possibility of making regulations applying to animals kept under intensive conditions on farms similar, for example, to those applicable to animals in transit. In other words, a sort of Factories Act for animals![45]

Another reviewer thought that while the "'natural food' line in the book may be less effective in practice than the cruelty line; the latter could really cause a stir especially bearing in mind the fact that it lends itself to pictorial treatment."[46] Two days after the publication of the first *Observer* article, another memo prepared MAFF officials for "the flood of questions and letters we must expect."[47] MAFF had two options:

> There is in fact a great deal to consider here and—unless the decision were to ride the storm—the course might in the end [be] to set up an independent committee of inquiry into the need for taking any action and the means which could be employed to take it.[48]

Another way of deflecting criticism might be to update existing regulations. In the case of John Dugdale's 1960 Bill on animal housing, MAFF officials had briefly considered extending the 1911 Act by granting cruelty inspectors right of entry on farms. However, the idea had been dropped as unpractical because it was thought that "rural benches" would be "particularly unwilling to convict"[49] potential delinquents. MAFF officials were also afraid of scaring off agricultural investors: "even rumours of legal standards to come" would make large poultry producers "hesitate before launching into vast enterprises because their capital costs would be raised."[50] Another minute therefore advocated maintaining MAFF's unofficial philosophy of equating farm animal welfare with productivity:

[44] TNA MAF 293/169 Minute W.C. Tame to Mr Hutchison (28.02.1964).
[45] TNA MAF 293/169 Minute W.C. Tame to Mr Hutchison (28.02.1964); Woods, "From Cruelty to Welfare," 18.
[46] TNA MAF 260/351 Minute J.A. Barrah to Mr Hensley (25.02.1964).
[47] TNA MAF 293/169 Minute N.J.P. Hutchison to Mr Virgo (03.03.1964).
[48] TNA MAF 293/169 Minute N.J.P. Hutchison to Mr Virgo (03.03.1964).
[49] TNA MAF 293/169 Minute N.J.P. Hutchison to Mr Virgo (03.03.1964).
[50] TNA MAF 293/169 Minute G.P. Jupe to Mr McPhail (03.03.1964).

> All our livestock are bred and reared for ultimate slaughter. ... There are some to whom this concept is in itself anathema, but it is their lot. Our concern is to ensure that man does not aggravate the essential bestiality of an animal's existence—or degrade himself—by imposing in its lifetime or at its death circumstances which cause avoidable suffering. ... The health and comfort of animals are measurable both veterinarily and humanely by the simple test of how they thrive: the efficiency of their metabolism. ... To justify control of any circumstances imposed by man on animals we need evidence either that it is physically cruel or that, because its effect is to cause an animal not to thrive.[51]

Initially, it seemed as though MAFF would indeed attempt to "stone-wall"[52] *Animal Machines*. On March 10, a departmental meeting confirmed that the

> general feeling, advanced most strongly by Animal Health Division, was that we should advise against any general enquiry: this was because the misgivings spring from an interpretation of animals' feelings which can only be a matter of individual opinion. It was recognized that some critics, like Ruth Harrison, a vegetarian [sic], have deep convictions about the use of animals for food at all.[53]

In a pre-written draft response to constituent enquiries, Conservative Minister of Agriculture Christopher Soames stated "that the Ministry does not accept many of [Ruth Harrison's] statements."[54] Although MAFF was taking Harrison's allegations seriously, "there is no evidence to show that these systems are generally cruel."[55] This line was in accordance with agro-industrial interests. Writing to Soames on March 17, the General Secretary of the Poultry and Egg Producers Association of Great Britain requested "that the Government's already declared policy ... that there is no evidence of cruelty, be re-affirmed in the clearest and most forthright terms."[56] In his draft speech for the Annual Dinner of the National Egg

[51] TNA MAF 293/169 Minute E.S. Virgo to Mr Hutchison (04.03.1964).

[52] TNA MAF 293/169 Minute N.J.P. Hutchison to Mr Virgo (03.03.1964).

[53] TNA MAF 260/351 Minute E.S. Virgo to D. Hadley (12.03.1964).

[54] TNA MAF 293/169 Draft Letter to Alfred Weirs, Esq. Darenth, Kent.

[55] TNA MAF 293/169 Draft Letter to Alfred Weirs, Esq. Darenth, Kent; TNA MAF 293/169 [Handwritten] 1st Draft: reply to debate on Mrs Joyce Butler's Motion 26/3, Poultry Husbandry.

[56] TNA MAF 293/169 T.J. Aley to Rt. Hon Christopher Soames (17.03.1964).

Packers' Association one day later, MAFF's Parliamentary Secretary toasted the "the British Egg Industry—that go-ahead problem child of the agricultural industry."[57] Briefly touching upon *Animal Machines*, he agreed that developing a "truly effective stunner for poultry" was desirable. However, MAFF would "not accept … that intensive farming methods as such [sic] involve unnecessary suffering."[58] While occasional management mistakes might occur, the same also held true for "more traditional forms of farming."[59] In comparison to pre-war farming, modern methods were luxurious. One might even draw an analogy to humans:

> I was just reflecting that we are all enjoying the benefits of intensive management at this very moment—warmth, ample food and drink, comfort, subdued lighting (or perhaps that will come later). Gentlemen, what more can one ask?[60]

MAFF's strategy of stonewalling *Animal Machines* underestimated the onslaught of public and political pressure for meaningful reform. Described by Karen Sayer, BBC One aired a 30-minute broadcast on "Farming: Animal Machines," and theatre plays and meetings across the country took up Harrison's criticism of intensive agriculture.[61] In parliament, Labour MPs Joyce Butler, John Rankin, and Fenner Brockway—who had supported Stephen Winsten's conscientious objection in 1916 (Chap. 2)—used *Animal Machines* to attack the already-struggling Conservative government under Alec Douglas-Home.[62]

Prominent animal protection organisations also became involved. Acknowledging *Animal Machines'* role in triggering its increased engagement with farm animal welfare,[63] the RSPCA requested a Departmental

[57] TNA MAF 260/351 Draft Speech for the Parliamentary Secretary at the Annual Dinner of the National Egg Packers' Association (18.03.1964), 1.

[58] TNA MAF 260/351 Draft Speech for the Parliamentary Secretary at the Annual Dinner of the National Egg Packers' Association (18.03.1964), 3

[59] TNA MAF 260/351 Draft Speech for the Parliamentary Secretary at the Annual Dinner of the National Egg Packers' Association (18.03.1964), 3.

[60] TNA MAF 260/351 Draft Speech for the Parliamentary Secretary at the Annual Dinner of the National Egg Packers' Association (18.03.1964), 3.

[61] Sayer, "Animal Machines," 486.

[62] Sayer, "Animal Machines," 487; TNA MAF 293/169 PQs Intensive Rearing Methods (23.03.1964).

[63] RSPCA Archives, IF/25/1 RSPCA Intensive Farming 2 of 2, RSPCA Annual Report for 1964, 254.

Committee "to inquire into the whole problem as affecting the intensive rearing and keeping of food animals in this country."[64] RSPCA representatives moreover sought a reintroduction of the failed 1960 Animals (Control of Intensified Methods of Food Production) Bill and lamented that inspectors "could not prosecute with any reasonable hope of success under the existing [1911] law."[65] The RSPCA Council also upgraded the prominence of farm animal welfare within the Society by resolving to set up a sub-committee on intensive methods of animal husbandry.[66]

Nine days after *Animal Machines'* publication, MAFF officials had no choice but to adapt their initial strategy. Although Minister Soames continued to publicly dispute "the facts as Mrs. Harrison sees them,"[67] Deputy Secretary Tame responded to RSPCA lobbying by conceding that "as long as anyone ... can start up these intensive systems without any kind of restriction on the conditions under which the animals are kept, we have no satisfactory answer to the humanitarians."[68] According to Tame, MAFF's only way to forestall "highly embarrassing recommendations" from an independent committee was to quickly "lay down statutory minimum standards for animals kept under intensive conditions."[69] Coupled with powers of entry or enforcement, such "standards need not be any more exacting than what is already good commercial practice."[70] Assessing the overall public mood, Permanent Secretary Sir John Winnifrith agreed that "it is a dead cert that there will be an intensive and persistent campaign to induce the Government to take action."[71] However, in Sir John's opinion,

[64] TNA MAF 260/351 Minute E.S. Virgo to Mr Hutchison (18.03.1964); RSPCA Archives, CM/55 1962-1966, Meeting of the Council, 19.03.1964, 4.

[65] TNA MAF 260/351 Minute E.S. Virgo to Mr Hutchison (18.03.1964); TNA MAF 293/169 John Hall to Mr Soames (12.03.1964).

[66] RSPCA Archives, CM/55, Meeting of the Council, 21.05.1964, 3; Meeting of the Council, 11.06.1964, 4; the sub-committee failed to gain influence until Ruth Harrison was elected to the RSPCA Council in 1969 and pushed for an ad hoc farm animals committee; RSPCA Archives, CM/58 Committee Minutes 1968-1970, 22.05.1969, 6; Meeting of the Council, 26.06.1969, 3.

[67] TNA MAF 260/351 Background Notes and Possible Supplementary Questions, Enclosed in: Minute ES Virgo to Mr Hutchison (18.03.1964).

[68] TNA MAF 121/267 Minute W.C. Tame to Mr Wall (19.03.1964).

[69] TNA MAF 121/267 Minute W.C. Tame to Mr Wall (19.03.1964).

[70] TNA MAF 121/267 Minute W.C. Tame to Mr Wall (19.03.1964).

[71] TNA MAF 121/267 Minute A.J.D. Winnifrith to Mr Evans (19.03.1964).

critics would be appeased only by a Departmental Committee examining the "cruelty aspect" of "concrete [sic] farming."[72]

After discussing the two options between March 19 and 20, Minister Soames agreed to appoint a "Departmental Committee with terms of reference confined to the issue of cruelty involved."[73] However, when senior MAFF officials discussed this option with Minister Soames during another meeting on March 25,[74] it became clear that a Departmental Committee would not be enough:

> there is the public aspect to consider. We have ... to be seen to be tackling [the problem]; and it may be very desirable to demonstrate that independent minds are being brought to bear. If everything were to be done within the Department there might be suspicions that we were covering up.[75]

Officials thus decided to follow a classic corporatist strategy by bringing representatives from all sides on board and crafting a palatable compromise.[76] It was advised to establish a Mixed Committee consisting of both MAFF and carefully chosen independent experts to "examine the conditions in which livestock are kept and managed under intensive husbandry methods and to recommend the standards that should obtain there."[77] It was, however, left unclear whether committee recommendations would be guidelines or "enforceable minima"[78] requiring legislation.

Following Soames' consent to the establishment of a Mixed Committee under an independent chairman, the search for committee members began.[79] Candidates for position of chairman were chosen not on the basis of their disciplinary expertise but on the basis of political experience and scientific standing. Suggestions included microbiologist Professor Ashley A. Miles of the Lister Institute; molecular biologist Professor Michael Swann of the University of Edinburgh; physiologist Sir Lindor Brown of the University of Oxford; and Lady Albemarle, who had recently chaired

[72] TNA MAF 121/267 Minute A.J.D. Winnifrith to Mr Evans (19.03.1964).
[73] TNA MAF 121/267 Minute M.D.M. Franklin to Mr Wall (20.03.1964).
[74] TNA MAF 121/267 Peter Pooley to Mr Franklin (25.03.1964), 1.
[75] TNA MAF 121/267 Minute R.G.R. Wall to Mr Franklin (24.03.1964), 1.
[76] Cox, Lowe, and Winter, "From State Direction to Self-Regulation," 475–490.
[77] TNA MAF 121/267 Minute R.G.R. Wall to Mr Franklin (24.03.1964), 2.
[78] TNA MAF 121/267 Minute R.G.R. Wall to Mr Franklin (24.03.1964), 2.
[79] TNA MAF 121/267 Peter Pooley to Mr Franklin (25.03.1964), 1.

a committee on youth and development in the community.[80] In the meantime, MAFF's "Parliamentary Secretaries should focus the attention of the Houses on the problem of standards, and shoot down the suggestions which we are not willing to investigate."[81]

Negotiations about the committee's size and membership were difficult. After the Home Office vetoed expanding the 1911 Act's rights of entry for cruelty inspections,[82] Minister Soames emerged as a driving force for the quick establishment of a committee. Having agreed to a *Daily Mirror* interview on "battery farming" on April 14, Soames was "anxious that, if at all possible we should forestall the *Daily Mirror* articles by getting out the news of the setting up of an Advisory Committee."[83] Unfortunately, the envisaged chairman cancelled shortly after the official announcement of the planned new Committee on April 20.[84] Following the suggestion of Chief Scientific Adviser Sir Solly Zuckerman, MAFF approached medical scientist Francis William Rogers Brambell, Professor of Zoology at Bangor in Wales.[85] Following Brambell's acceptance of chairmanship in mid-May, the other committee members were appointed by May 29, 1964.[86]

The resulting Technical Committee to Enquire into the Welfare of Animals Kept Under Intensive Livestock Husbandry Systems started work in July 1964.[87] At first glance, the dominance of MAFF-selected specialists did not promise radical reform. Amongst the minority of "enlightened laymen"[88] was radio and television personality Lady Isobel Barnett, whose main virtues according to Permanent Secretary Winnifrith were that "many millions" knew her and "that she is most unlikely to be cranky."[89]

[80] TNA MAF 121/267 Peter Pooley to Mr Franklin (25.03.1964), 1; TNA MAF 121/267 Minute R.G.R. Wall to Mr Franklin (24.03.1964), 2; TNA MAF 121/267 P. Humphreys-Davies (07.04.1964).

[81] TNA MAF 121/267 Peter Pooley to Mr Franklin (25.03.1964), 2.

[82] TNA MAF 121/267 J.C. Green to N.J.P. Hutchison (14.04.1964).

[83] TNA MAF 121/267 Minute Peter Pooley to Mr Wall (14.04.1964).

[84] "House of Commons", *Times*, 12.05.1964, 16.

[85] TNA MAF 121/267 Peter Pooley to Mr Wall (22.04.1964); A.J.D. Winnifrith to Mr Wall (29.04.1964).

[86] TNA MAF 121/267 W.C. Tame to Mr Cannel (13.05.1964); C.R. Cann to Mr Wall (29.05.1964).

[87] TNA MAF 121/267 C.R. Cann to Mr Wall (29.05.1964); TNA MAF 121/268 Committee on Intensive Husbandry Systems, Minutes of First Meeting (16.07.1964).

[88] MAF 121/267 R.G.R. Wall to Mr Virgo (26.05.1964).

[89] TNA MAF 121/267 A.J.D. Winnifrith to Parliamentary Secretary (10.06.1964), 2.

Ruth Harrison was not part of the committee.[90] Despite the inclusion of the term "welfare" in its official name,[91] Brambell Committee members referred to themselves as the "Committee of Enquiry into Intensive Livestock Husbandry Systems."[92]

For Ruth Harrison, the fact that MAFF had been forced to announce an independent committee within little over a month after the publication of *Animal Machines* was a major victory.[93] Both the timing and promotion of the publication of *Animal Machines* had been perfect. However, its author now faced the decision between whether to retire from campaigning or to continue. Ruth Harrison chose the latter course.

The weeks and months following *Animal Machines'* publication had been a veritable crash course in public relations. During a four-hour rally at Trafalgar Square, which had been organised by the closely affiliated Animal Machines Action Group and co-financed by the RSPCA, Ruth Harrison spoke to 200 people, who had just delivered a petition with 250,000 signatures to end 'factory farms' to MAFF.[94] On May 12, Ruth Harrison met four MPs to discuss 'factory farming' in the House of Commons.[95]

She also gained significant media experience. On March 11—two days after the publication of *Animal Machines*—she appeared as a guest on the BBC's radio programme *On Your Farm*. The show was presented by farmer Bryan Platt, who had also invited Dr John Williams, director of the Animal Feeds Division of Ful-O-Pep Feeds, to discuss *Animal Machines*.[96] From the beginning, both Platt and Williams attempted to downplay Harrison's allegations. Using her first name, Williams in particular tried to portray Harrison as an idealistic, yet overly emotional housewife, whose amateur research did not bear up to 'hard' scientific scrutiny:

[90] This is in contrast to claims by Donald M. Broom, "Ruth Harrison's Later Writings and Animal Welfare Work," *Animal Machines—New Edition* (Wallingford and Boston CABI, 2013), 21.

[91] Woods, "Cruelty to Welfare," 18.

[92] TNA MAF 121/268.

[93] A.H. Sykes, "The fatted fowl", *Guardian*, 11.08.1964, 6.

[94] Sayer, "Animal Machines," 486; "Factory farms protest", *Observer*, 25.04.1965, 4; RSPCA Archives, CM 55 1962-1966, Meeting of the Council, 15.04.1965, 3.

[95] "MPs see author about factory farming", *Daily Mail* (12.05.1964), 12; John Winter, "Farm Mail", *Daily Mail* (13.05.1964), 16.

[96] TNA MAF 260/351 Transcript from a Telediphone Recording from Talks/ General Division—Sound: On Your Farm (12:30; 11.03.1964).

> I would like, Ruth, right at the outset to say that I don't wish to make a point to attack you. ... But I'm sure also that you're—in putting the thoughts which you have on paper—you are extremely motivated by the highest ideals and ... indeed haven't tried to write something which is a piece of sensationalism. ... Your conclusions, I think, are somewhat illogical in that they are surrounded with an aura of sentimentality and mixed opinions.[97]

Following this, Harrison was allowed to present her position in one-sentence statements before either Platt or Williams weighed in and demanded evidence or corrected alleged misconceptions. Platt defended the equation of welfare with animal thrift before remembering that he was supposed to be a neutral host: "I'm sorry, I hope we're not being unfair to you here at all. I am very much involved as a farmer and it's difficult to be impartial as a chairman at the same time."[98] Following this, the discussion shifted to the stunning of animals and the safety of intensively farmed food. Williams referred to lack of evidence, isolated problems, and sentimental urban anthropomorphism to counter Harrison. Only in the case of antibiotics did he admit that "there has been wide abuse. ... No-one would disagree with this I think. You can say that this is a majority practice, oddly enough, as opposed to a minority. I personally [am] completely opposed to the ... the widespread use."[99]

The experience of being treated in a condescending manner and seeing her allegations trivialised was not unique to *On Your Farm*. In 1975, Ruth Harrison remembered the radio debate and others like it during an interview published in *The Vegetarian* and *Observer Colour Magazine*.[100] She was particularly annoyed by a debate with an official of the British Broiler Association:

> I was tremendously unsophisticated and made my main points in what was supposed to be a pre-broadcast run through. The PR man took them down, and fed through the answers to his boss. The producer said there was no

[97] TNA MAF 260/351 Transcript from a Telephone Recording from Talks/ General Division—Sound: On Your Farm (12:30; 11.03.1964), 7.

[98] TNA MAF 260/351 Transcript from a Telephone Recording from Talks/ General Division—Sound: On Your Farm (12:30; 11.03.1964), 9.

[99] TNA MAF 260/351 Transcript from a Telephone Recording from Talks/ General Division—Sound: On Your Farm (12:30; 11.03.1964), 12.

[100] Kendall, "Ruth and the Ruthless"; the interview also appeared in the *Observer Colour Magazine*.

time for my opponent's main points to be put. When it went over the air they represented me as an emotional housewife, not backed up with facts. I've learnt a lot through bitter experience.[101]

Ruth Harrison was also invited to give evidence to the Brambell Committee. Accompanied by her husband, she appeared in front of the committee on November 20, 1964.[102] By this time, committee members had already toured several intensive production facilities and assessed the debeaking of birds, high humidity houses for pigs, floor feeding, and slatted floors.[103] Most importantly, they had heard a paper on animal welfare and affective states given by committee member and Harrison acquaintance William Homan Thorpe, which discussed the "sophisticated social life"[104] of all animals held under intensive positions. Building on the growing body of ethological research on animal cognition (Chap. 4), Thorpe thought it wrong to limit farm animal welfare considerations to the absence of physical pain and lack of productivity: "Animals were more intelligent and their behaviour patterns more complex than often appeared when they were kept under domesticated conditions."[105] As a consequence, animals' welfare could also be impacted by psychological stress. Later appearing as an annexe to the official Brambell Report, Thorpe's paper specifically objected to: "conditions which lead to physical deformity and to highly abnormal nutritional physiology" and "conditions which completely suppress all or nearly all the natural, instinctive urges and behaviour patterns characteristic of actions appropriate to the high degree of social organisation."[106]

In her statement to the Brambell Committee, Ruth Harrison made it clear that she objected to intensive husbandry in general. She was

[101] Quoted according to Kendall, "Ruth and the ruthless," 2.

[102] TNA MAF 121/268 Committee of Enquiry into Intensive Husbandry Systems. Minutes of meeting, interview Ruth Harrison (20.11.1964), 1.

[103] TNA MAF 121/268 Minutes of Meeting held at Queen's Hotel, Leeds (23.09.1964), 3; TNA MAF 121/268 Minutes of Meeting held at "Great Danes" Inn, Hollingbourne (30.09.1964); TNA MAF 121/268 Minutes of Meeting held George Hotel, Nottingham (22.10.1964).

[104] TNA MAF 121/268 Minutes of Meeting held George Hotel, Nottingham (22.10.1964), 2.

[105] TNA MAF 121/268 Minutes of Meeting held George Hotel, Nottingham (22.10.1964), 2.

[106] *Report of the Technical Committee to Enquire into the Welfare of Animals kept under Intensive Livestock Systems* (London: HMSO, 1965), 79.

particularly concerned about "any system of husbandry which she believed to have gone so far in intensivism as to make it impossible to maintain the health of the stock without resorting to the heavy use of drugs..."[107] Harrison noted that "the traditional farmer might cause suffering through ignorance or inefficiency, but this was different from using sophisticated techniques which resulted in misery for animals."[108] In particular, Harrison objected to intensive veal crate and barley beef production systems, fully slatted sties, debeaking, dim lighting, battery cages, sweat houses, and restricting freedom of movement. However, in an important sign of things to come, "Mrs. Harrison's objections of principle were tempered with an appreciation of economic necessities."[109] She did not oppose slatted floors in dunging areas, more space for poultry within existing systems, or floor feeding for pigs. To promote good welfare, Harrison advocated amending the 1911 Act to allow enforcement "by any member of the public rather than by regulations enforced by the [MAFF]."[110] RSPCA inspectors should be allowed to access "farms and other premises without resorting to the police."[111] Most importantly, further refinements to existing systems should have to comply with legally defined welfare standards guaranteeing "freedom of movement, proper diet, adequate lighting and health."[112]

This appreciation of economic and political necessities was also evident in the Ruth Harrison Advisory Group's submission to the Brambell Committee. Instead of following more radical calls for a complete ban on 'factory farming,' the group advocated mandatory training and

[107] TNA MAF 121/268 Committee of Enquiry into Intensive Husbandry Systems. Minutes of meeting, interview Ruth Harrison (20.11.1964), 1.

[108] TNA MAF 121/268 Committee of Enquiry into Intensive Husbandry Systems. Minutes of meeting, interview Ruth Harrison (20.11.1964), 1.

[109] TNA MAF 121/268 Committee of Enquiry into Intensive Husbandry Systems. Minutes of meeting, interview Ruth Harrison (20.11.1964), 2.

[110] Ruth Harrison, "Letter to the Editor viz. Factory farming is here to stay", *Guardian*, 02.01.1965, 3; Harrison's positions in front of the Brambell Committee were nearly identical to those expressed by the RSPCA during a further session on the same day; TNA MAF 121/268 Committee of Enquiry into Intensive Husbandry Systems. Minutes of meeting, interview RSPCA (20.11.1964).

[111] TNA MAF 121/268 Committee of Enquiry into Intensive Husbandry Systems. Minutes of meeting, interview Ruth Harrison (20.11.1964), 3.

[112] TNA MAF 121/268 Committee of Enquiry into Intensive Husbandry Systems. Minutes of meeting, interview Ruth Harrison (20.11.1964), 3.

examinations for "all staff responsible for farm birds and beasts,"[113] farm licences reflecting a fixed ratio of livestock to acreage, statutory standards of animal treatment, and required access to open ground for all stock of appropriate age. Further suggestions were the establishment of a welfare research centre, restricting antibiotics to therapeutic uses, establishing a code of nutritional practice, and restricting non-compliant meat imports.[114]

Aware that it would be futile to challenge intensive animal husbandry's very existence, Harrison and her close allies adopted a twofold strategy of pushing for gradualist statutory reforms in what Angela Cassidy has termed the "backstage"[115] of official circles while continuing to promote in public a wholesale consumer revolt against intensive farming. As a 1965 letter to the *Guardian* makes clear, Ruth Harrison ultimately trusted well-informed consumers more than official standards to generate meaningful welfare improvements:

> If the shopper is educated as to what her [sic] choice means, then she can discriminate. If we identify and reject the 'factory produced' product, whether from at home or abroad, then we can influence the use of these methods. ... What we can hope, with Rachel Carson, is for 'a consumers' revolt of such proportions' as will reject these inhumane methods, and count the cost in the true, humane sense.[116]

Over the next three decades, Harrison would relentlessly pursue this dual strategy of pushing hard for any possible improvement within the existing system of intensive farming while publicly opposing it in principle. As the author of *Animal Machines*, she was able to present herself as a useful intermediary, who could develop compromise policies with officials and moderate organisations like the RSPCA yet also had the ear of more radical activists. Maintaining this outsider status and continuously blurring the 'front'- and 'backstage' of politics exposed her to criticism from both sides. According to Harrison, "you're going to offend people whatever

[113] "Charter suggested for farm animals' condition", *Guardian*, 25.11.1965, 13.

[114] TNA MAF 121/268 Committee of Enquiry into Intensive Husbandry Systems. Minutes of meeting, interview Ruth Harrison (20.11.1964), 2; for parallel debates on antibiotic restrictions see Kirchhelle, *Pyrrhic Progress*, Chapter 7.

[115] Cassidy, *Vermin*, 205.

[116] Ruth Harrison, "Letter to the Editor viz. Factory farming is here to stay", *Guardian*, 02.01.1965, 8.

you do ... You have almost to be a lone fighter."[117] Without firm allies in either camp, Harrison's main source of moral and public authority continued to be *Animal Machines*. However, given the rapid influx of new voices into the field of animal welfare, even this strong source of public authority was not limitless.

Inspired by the success of *Animal Machines*, other authors began to plough the field of animal welfare. In 1965, full-time writer Elspeth Huxley published *Brave New Victuals*. Drawing on her family connections and previous writing on wildlife conservation,[118] Huxley aimed for a balanced account of intensive farming somewhere between the "land of crackpots, cranks and the lunatic fringe" and "scientific fact."[119] Although she did not reference it, Huxley tackled most of *Animal Machines'* core themes: intensive agriculture, environmentalism, veal production, battery eggs, broiler hens, and sweat boxes for pigs. However, unlike Harrison, Huxley did not pass judgement. Describing the tens of thousands of battery hens held at Eastwood farms, she asked: "Is all this cruel? Certainly it is 'unnatural'. But so are all forms of farming."[120] In the case of Northern Irish pig sweat boxes, she could not find any cruelty beyond that of other intensive methods.[121] Regarding veal calves, her findings were also noncommittal. While early uses of Dutch systems had resulted in excess mortality, modifications like slatted crates and dim lighting had solved many problems:

> A calf penned into a narrow crate is no worse off, better off in most cases, than a calf tied up night and day in a dark, dirty corner of a draughty, fly-infested shed on a traditional farm (...), they are not, by reasons of their haemoglobin levels, in any discomfort, let alone pain. (...). The crated calves

[117] Quoted according to Kendall, "Ruth and the Ruthless," 21.

[118] Elspeth Huxley had already published numerous nonfiction and fiction books. Many focused on her previous life in Kenya and issues surrounding wildlife conservation. She was married to Gervas Huxley, cousin of Aldous and Julian Huxley, and maintained extensive correspondence with Julian Huxley on ethology, conservation, and environmentalism; Rice University, Julian Sorrell Huxley Papers, Correspondence Elspeth Huxley.

[119] Elspeth Huxley, *Brave New Victuals. Are We All Being Slowly Poisoned? A Terrifying Enquiry Into The Techniques of Modern Food Production* (London: Panther Books [1965] 1967), 10.

[120] Huxley, *Brave New Victuals*, 22.

[121] Huxley, *Brave New Victuals*, 45.

I saw did not look unhappy; glossy of coat, bright of eye, well ventilated, clean, free of flies.[122]

Although she criticised liver problems resulting from barley beef production, Huxley emphasised that scientific judgement on most intensive systems remained out.[123] Ethological research showed that higher herd densities and reduced space could trigger stress, and new rules might eventually be necessary "to catch up with our techniques."[124] However, scientists might also be able to "dislodge"[125] stress genes. Leaving it to readers to decide whether intensive farming was cruel, Huxley instead devoted the majority of her book to the quantifiable health and environmental hazards of new weedkillers, pesticides, hormones, additives, and antibiotics in the food sector.[126] According to Huxley, contrasting a romanticised world of small farms with large intensive operations was a naive juxtaposition. Change was inevitable. Regardless of whether one wanted to preserve the countryside or protect animals, the realities of global population growth meant that "a lot of animals must stay indoors."[127] But which principles should guide the design of these production facilities?

The Brambell Committee published its much-anticipated report on intensive livestock husbandry shortly after Huxley's book in December 1965. The report reflected months of intensive lobbying by all involved parties. Both the NFU and the Animal Health Trust had opposed mandatory husbandry standards.[128] Interested in promoting state-sponsored preventive medicine,[129] the BVA upheld "thrift" as a moderately "accurate index of welfare"[130] but called for an advisory veterinary health

[122] Huxley, *Brave New Victuals*, 53-54.

[123] Huxley, *Brave New Victuals*, 20.

[124] Huxley, *Brave New Victuals*, 70.

[125] Huxley, *Brave New Victuals*, 69.

[126] Huxley, *Brave New Victuals*, 15-17, 34-40, 71-107.

[127] Huxley, *Brave New Victuals*, 112.

[128] TNA MAF 121/268 Committee of Enquiry into Intensive Husbandry Systems. Minutes of meeting, interview Animal Health Trust (03.12.1964); TNA MAF 121/268 Committee of Enquiry into Intensive Livestock Husbandry Systems. Minutes of meeting, interview NFU (18.12.1964); the Trust approved of health records oversight by veterinary inspectors.

[129] Abigail Woods, "Is Prevention Better Than Cure? The Rise and Fall of Veterinary Preventive Medicine, C. 1950-1980," *Social History of Medicine* 26/1 (2012), 113–131.

[130] TNA MAF 121/268 Committee of Enquiry into Intensive Husbandry Systems. Minutes of meeting, interview BVA (26.01.1965), 1.

service to promote and oversee good animal welfare. Although it also failed to advocate statutory regulations, the Royal College of Veterinary Surgeons (RCVS) was more explicit in criticising practices such as "deprivation of light, unnatural restraint of movement, the use of diets intended to induce a pathological state," and unnecessary mutilations.[131] Animal protection and consumer organisations called for statutory reform. According to the UFAW, it was wrong to claim that "thrift was automatic proof of an acceptable level of welfare."[132] The National Federation of Women's Institutes was concerned about the rise of the "non-rural business man whose concern was to extract the highest possible profit from livestock without any concern for well-being."[133] Not opposed to intensive farming in principle, the Federation criticised "close confinement" and "unnecessary use of drugs or antibiotics [sic]."[134] The RSPCA called for firmer regulation and subsequently launched publicity campaigns for free-range eggs and against government subsidies for new intensive broiler plants in Wales.[135] The most outspoken condemnation of intensive farming came from the Humane Farming Campaign, which accused "exponents of intensivism" of "running counter to accepted Christian beliefs."[136] There were "many biblical warnings against such behaviour": "Mankind would ultimately suffer as a result of the debasement of the quality of life ... The wrath of God would be called down on our civilisation, as it was on the Egyptians."[137]

[131] TNA MAF 121/268 Committee of Enquiry into Intensive Husbandry Systems. Minutes of meeting, interview RCVS (12.02.1965), 2.

[132] TNA MAF 121/268 Committee of Enquiry into Intensive Husbandry Systems. Minutes of meeting, interview UFAW (26.01.1965), 3.

[133] TNA MAF 121/268 Committee of Enquiry into Intensive Husbandry Systems. Minutes of meeting, interview National Federation of Women's institutes (12.02.1965), 1.

[134] TNA MAF 121/268 Committee of Enquiry into Intensive Husbandry Systems. Minutes of meeting, interview National Federation of Women's institutes (12.02.1965), 2.

[135] TNA MAF 121/268 Committee of Enquiry into Intensive Husbandry Systems. Minutes of meeting, interview RSPCA (20.11.1964); RSPCA Archives, CM/55 1962-1966, Meeting of the Council, 11.06.1964, 3-4; Meeting of the Council, 17.02.1966, 3; Meeting of the Council, 17.03.1966, 3.

[136] TNA MAF 121/268 Committee of Enquiry into Intensive Husbandry Systems. Minutes of meeting, interview Humane Farming Campaign (26.01.1965), 1.

[137] TNA MAF 121/268 Committee of Enquiry into Intensive Husbandry Systems. Minutes of meeting, interview Humane Farming Campaign (26.01.1965), 1.

Navigating between predictions of agricultural bankruptcy and divine wrath, the main Brambell Report recommended promoting more behavioural research on animal welfare, improving the education of stockmen, and installing a standing Advisory Committee to advise Ministers on farm animal welfare. Members of the Advisory Committee should be appointed for their own qualifications and not as representatives of particular interests.[138] Although it made concrete recommendations regarding space, lighting, diets, ventilation, and flooring and also criticised practices such as debeaking and tethering,[139] the Brambell Committee did not advocate amending the 1911 Act. Instead, it called for a "new and brief animal welfare act."[140] Handed to MAFF on October 25 and published on December 2, 1965,[141] the report's most lasting contribution was to call for an expansion of the definition of animal suffering—or negative welfare—from physical pain and mental cruelty as defined by the 1911 Act to encompass discomfort, stress and pain: welfare was a "wide term that embraces both the physical and mental well-being of animals."[142] Building on Thorpe's recommendations, the report stated that an "animal should at least have sufficient freedom of movement to be able without difficulty, to turn round, groom itself, get up, lie down and stretch its limbs."[143] What a positive state of well-being might look like beyond the absence of negative welfare factors like stress was not discussed.

In their initial assessment of the Brambell Report, MAFF officials were surprised by how cheap it would be to implement many of its recommendations.[144] The major concern was how to deal with cheaper livestock imports from countries not adhering to Brambell standards: "the foreigner

[138] TNA MAF 121/268 Committee of Enquiry into Intensive Husbandry Systems. Minutes of meeting (13-15.07.1965), 2.

[139] *Report of the Technical Committee to Enquire into the Welfare of Animals kept under Intensive Livestock Systems* (London: HMSO, 1965), 63-65.

[140] TNA MAF 121/268 Committee of Enquiry into Intensive Husbandry Systems. Minutes of meeting (27.-28.07.1965), 2.

[141] TNA MAF 369/32 Animal Health Division II: Brambell Committee. Meeting of Officials on 18th November. Notes for Mr Humphreys-Davies (17.11.1965), 1.

[142] *Report of the Technical Committee to Enquire into the Welfare of Animals*, 9; Woods, "Cruelty to Welfare," 19; this definition was partly inspired by the recent Littlewood Report on Experiments on Animals.

[143] *Report of the Technical Committee to Enquire into the Welfare of Animals*, 13.

[144] TNA MAF 369/32 Minute A.J.D. Winnifrith to Minister (24.11.1965).

might be enabled to get a bigger share of the UK market."[145] MAFF Deputy Secretary Peter Humphreys-Davies noted:

> the [Brambell] Committee have struck a reasonable balance between the agriculture and food interests on the one hand and the sentimentalists on the other. The producers, for instance, have only lost out completely on white veal and the pig seat-houses. The battery system only requires to be modified and not prohibited, ... Debeaking was on the way out anyway. ... The worst of these, perhaps, is the proposal that the same standards should be applied to imports: ... The recommendation about the tethering of beef cattle (and by implication dairy cows) in winter is also clearly unacceptable as it stands. ... The cost considerations ... are to my mind far less formidable: given a reasonable transitional period, and perhaps some help in the Annual Review, the extra cost ought to be absorbed easily enough in course of time.[146]

Almost two years after the publication of *Animal Machines*, public pressure for immediate welfare reform had, however, diminished. With Britain struggling to maintain its balance-of-payments,[147] MAFF officials were unwilling to allow welfare reforms to jeopardise agricultural efficiency. Meanwhile, the Home Office insisted that the 1911 Act was a sufficient safeguard for animal protection and opposed including the word "stress"[148] in a new welfare bill. Despite positive continental reactions to the Brambell Report,[149] British regulators decided to either ignore or

[145] TNA MAF 369/32 Fatstock Policy Division, Report of the Brambell Committee (15.11.1965), 1; TNA MAF 369/32 Animal Health Division II: Brambell Committee. Meeting of Officials on 18th November. Notes for Mr Humphreys-Davies (17.11.1965).

[146] TNA MAF 369/32 Minute P. Humphreys-Davies to Secretary (24.11.1965).

[147] Martin, *The Development of Modern Agriculture*, 88-91; fears of lacking efficiency vis-à-vis EEC competitors were repeatedly mentioned by the NFU; TNA MAF 369/77 Farm Animal Welfare Legislative Proposals. Note of a meeting with the NFU held on 16.03.1967 (01.05.1967), 1.

[148] TNA MAF 369/47 Draft Minute, Enclosed in, Minute G.O. Lace to Mr Humphreys Davies (31.08.1966)

[149] Bayerisches Hauptstaatsarchiv (HSTA)—MInn 87782 III-5594/2-2/66 Der BMELF an die für das Veterinärwesen zuständigen obersten Landesbehörden (11.08.1966).

weaken many of its recommendations. Legislation to improve the education of stockmen was considered impracticable.[150] In the case of the committee's space recommendations, MAFF officials contradicted earlier assessments and warned, "if implemented, Brambell must marginally, but perhaps overall imperceptibly, increase the cost of food; will raise costs or reduce output for some producers far from imperceptibly; and will check the momentum of increased efficiency."[151] According to Assistant-Secretary G.O. Lace, it was impossible to "see how Ministers can decide, before consulting the trade on fairly detailed proposals, either to what extent to modify Brambell, or where to place the cost of it."[152]

MAFF consultations with industry and welfare interests lasted for another year and further weakened many Brambell recommendations.[153] In its original comment on the Brambell Report, MAFF had already declared that it "could not impose detailed statutory standards at the present time."[154] Codes of practice would be drawn up, and state veterinarians would provide advice to farmers on these voluntary codes. Like the Highway Code, non-compliance would not be an offence but could be used to establish culpability in cruelty prosecutions.[155] The new welfare bill would not make sweeping reforms but would allow Ministers of Agriculture to make regulations on a case-by-case basis. A major Brambell recommendation to remain unchallenged was the establishment of a Farm Animal Welfare Advisory Committee (FAWAC). In August 1966, MAFF publicly agreed to set up an advisory committee on animal welfare.[156] However, the extent of the committee's power was contested. Officials thought it best not to let FAWAC re-examine systems that had "passed the Brambell Test" and saw "perhaps some nuisance, in the Committee being statutory."[157] Following drawn-out discussions, MAFF decided to estab-

[150] TNA MAF 369/47 G.O. Lace, "Note for Minister. Brambell Committee Report—Possible Courses of Action" (17.05.1966), 2.

[151] TNA MAF 369/47 G.O. Lace/Mr Hensley, "Brambell Report" (undated), 17; see also: 1-12.

[152] TNA MAF 369/47 G.O. Lace/Mr Hensley, "Brambell Report" (undated), 17.

[153] TNA MAF 369/77 Press Release MAFF (02.11.1967—1)

[154] TNA MAF 369/272 Joint Announcement by the Agricultural Departments in the United Kingdom MAFF (05.08.1966), 2.

[155] TNA MAF 369/272 Joint Announcement by the Agricultural Departments in the United Kingdom MAFF (05.08.1966), 2.

[156] TNA MAF 369/272 Joint Announcement by the Agricultural Departments in the United Kingdom MAFF (05.08.1966), 2.

[157] TNA MAF 369/47 G.O. Lace/Mr Hensley, "Brambell Report" (undated), 4.

lish FAWAC as a non-statutory, standing body "to advise ... on matters pertaining to the welfare of farm animals."[158] The committee was to have eight to ten "general members," who represented different interests but were not "nominees" of "particular interests."[159] FAWAC should comprise amongst others an animal behaviourist, a geneticist, and "one or two" zoologists and "independent veterinarians."[160] It was not necessary for the chairperson to have a "particular field of knowledge" but to be "strong-minded, independent and able to manage what could be a difficult committee."[161]

Despite her disappointment about the abandonment of many Brambell principles, Ruth Harrison quickly recognised that FAWAC offered an opportunity to influence the future development of British welfare regulations. Starting in August 1966, she used her status as a prominent yet moderate 'outsider' with semi-expert credentials to lobby MAFF for FAWAC membership. Having met Ruth Harrison for lunch on August 18, MAFF official D. Evans informed Assistant-Secretary G.O. Lace:

> [Ruth Harrison] was very interested in the Advisory Committee, and fished for a long time for me to say something about it. In the end she asked me direct whether I thought she would be invited to be a member; I said I honestly did not know, ... I supposed she might be regarded as eligible, but from my understanding of matters there might be a good deal of opposition from trade interests and others if she were invited. She told me that she would probably serve if asked, though she was a little worried about being 'gagged.'[162]

According to Evans, "Mrs. Harrison was at pains to impress on me that she regards herself as a 'moderate' in the Animal Welfare Movement."[163]

Not all officials were enthusiastic about appointing Ruth Harrison to FAWAC. Reviewing the list of potential candidates, MAFF Deputy Secretary Peter Humphreys-Davies recommended dropping both Ruth Harrison and her co-nominee the Dean of Llandaff.[164] MAFF

[158] TNA MAF 369/47 Minute P. Humphreys-Davies to Secretary (07.09.1966), 2.
[159] TNA MAF 369/47 Minute P. Humphreys-Davies to Secretary (07.09.1966), 2.
[160] TNA MAF 369/47 Minute P. Humphreys-Davies to Secretary (07.09.1966), 2.
[161] TNA MAF 369/47 Minute P. Humphreys-Davies to Secretary (07.09.1966), 2.
[162] TNA MAF 369/47 Minute D. Evans to Mr Lace (19.08.1966).
[163] TNA MAF 369/47 Minute D. Evans to Mr Lace (19.08.1966).
[164] TNA MAF 369/47 Minute P. Humphreys-Davies to Mr Hensley (11.01.1967).

under-secretary C.H.M. Wilcox favoured nominating former Brambell member Lady Isobel Barnett alongside Elspeth Huxley—on the basis of her 'balanced' *Brave New Victuals*—and excluding Ruth Harrison. However, he could "appreciate that there are no doubt strong political arguments for taking the latter."[165] By contrast, Under-Secretary of State for Scotland Lord Hughes, "when told that Mrs. Harrison was on the list, thought her better on than off."[166] This opinion was shared by G.O. Lace: "I gather that if we want anyone from outside the established societies she would indeed be the most moderate of the well-known people."[167] Meanwhile, MAFF official D. Evans indirectly pushed for Ruth Harrison by casting doubt on Elspeth Huxley's commitment to animal welfare:

> I think there is no doubt about her impartiality (though I am not sure that this is what is wanted in a 'welfare' member) but in my dealings with her when the Brambell Committee was sitting I gained the impression that she thought of her work on this subject simply as a professional journalist would do and had no lasting interest in it.[168]

All the while, Ruth Harrison continued to actively push for FAWAC membership. Phoning MAFF on September 14, 1966, "she told [Evans] she had given a lot of thought to the [Minister's proposed action on the Brambell Report] and had come to the conclusion that our proposals were much better than she originally thought."[169] Again stressing her credentials as a moderate, "she said she was thinking of writing to the Minister to offer her support."[170] In the course of the conversation, Evans once again "got the impression that Mrs. Harrison is very anxious to be on the Standing Committee."[171] Referring to Ruth Harrison's recent statements and "varying accounts" of Lady Isobel Barnett's performance on the Brambell Committee, another MAFF minute noted, "[Harrison] has taken a balanced and objective view and I think on the whole she should be invited."[172]

[165] TNA MAF 369/47 Minute C.H.M. Wilcox to W.E. Jones (09.09.1966).
[166] TNA MAF 369/47 Minute W.E. Jones to C.H.M. Wilcox (13.09.1966), 1.
[167] TNA MAF 369/47 Minute G.O. Lace to Mr Hensley (24.08.1966).
[168] TNA MAF 369/47 Minute D. Evans to Mr Hensley (15.09.1966).
[169] TNA MAF 369/47 Minute D. Evans to Mrs Avery (15.09.1966).
[170] TNA MAF 369/47 Minute D. Evans to Mrs Avery (15.09.1966).
[171] TNA MAF 369/47 Minute D. Evans to Mrs Avery (15.09.1966).
[172] TNA MAF 369/47 Minute P. Humphreys-Davies to Secretary (07.09.1966), 2.

Clearly designed to aid Harrison's FAWAC membership campaign, the Ruth Harrison Advisory Group submitted extremely moderate comments on MAFF's proposed animal welfare reforms in November 1966:[173]

> The Minister's Proposals could be an excellent framework within which a solution can be found for the ultimate welfare of food animals, further, they are sufficiently flexible to allow for further change in the light of increasing knowledge and public concern.[174]

In contrast to strong criticism from groups like the National Campaign for the Abolition of Factory Farming, the Humane Farming Campaign, and the RSPCA,[175] the Advisory Group merely called for more detailed voluntary codes for indoor and outdoor farming "after the manner of the advisory pamphlets issued by the Ministries since the last war."[176] It was hoped that the Brambell recommendations would not be interpreted as absolute standards but "as a minimum, and a point of departure."[177] Regarding welfare controls of ca. 360,000 agricultural holdings by ca. 600 veterinary surgeons and staff, the Advisory Group suggested that the Minister allow "any person or organisation, with the backing of a veterinary surgeon, to take action where this is necessary."[178] The Advisory Group's strongest criticism centred on the postponement of regulatory action until further scientific assessments had been completed:

> Whilst we understand the Ministers' reluctance to involve producers in capital expenditure ..., we are apprehensive that this could become a ground for

[173]TNA MAF 369/75 Minute G.O. Lace to Mr Hensley (16.11.1966).

[174]TNA MAF 369/75 Ruth Harrison Advisory Group, "Comment on the Minister's Proposals for Legislation following on the Brambell Report" (08.11.1966), enclosed in: Minute G.O. Lace to Mr Hensley (16.11.1966), 1.

[175]TNA MAF 369/75 Humane Farming Campaign—Farm Animal Welfare—Proposals for Legislation; TNA MAF 369/75 Lucy Newman to MAFF (16.11.1966); TNA MAF 369/77 RSPCA to HB Fawcett (12.09.1967).

[176]TNA MAF 369/75 Ruth Harrison Advisory Group, "Comment on the Minister's Proposals for Legislation following on the Brambell Report" (08.11.1966), enclosed in: Minute G.O. Lace to Mr Hensley (16.11.1966), 1.

[177]TNA MAF 369/75 Ruth Harrison Advisory Group, "Comment on the Minister's Proposals for Legislation following on the Brambell Report" (08.11.1966), enclosed in: Minute G.O. Lace to Mr Hensley (16.11.1966), 2.

[178]TNA MAF 369/75 Ruth Harrison Advisory Group, "Comment on the Minister's Proposals for Legislation following on the Brambell Report" (08.11.1966), enclosed in: Minute G.O. Lace to Mr Hensley (16.11.1966), 3.

> virtually indefinite postponement of action. ... If the Ministers incorporated the Brambell Committee recommendations into the Code of Practice, making each one statutory as soon as they felt they had sufficient evidence and support to do so, and if this were backed up by strong enough powers of prosecution, ... then we would welcome their proposals as being a very real step forward in the welfare of farm animals.[179]

Voicing similar demands, a group of well-known activists and experts, including Ruth Harrison, William Homan Thorpe, and Soil Association founder Eve Balfour, published a letter in the *Times* in May 1967:

> Only two steps are needed ..., first for the Minister of Agriculture to have [the Brambell] recommendations incorporated without any further modification in the proposed voluntary Code of Practice contemplated in his prospective Bill, and secondly for the Government to make time for the Bill to be passed through Parliament.[180]

New regulations and codes should enact basic Brambell recommendations: an animal should have freedom of movement, be able to turn round, groom itself, get up, lie down, and stretch its limbs; diets should be designed to maintain animals' full health and vigour; adequate illumination should be available for the proper and routine inspection of all animals. These recommendations "were so fundamental that one feels surprised that the necessity ever arose for them to be made."[181]

The letter failed to impact official decision-making. In August 1967, MAFF circulated new draft provisions for the upcoming welfare law. Little had changed. The usually moderate RSPCA criticised the ongoing absence of statutory regulations except for proposed bans on the bleeding of calves and tail-docking of cattle alongside regulations for adequate lighting and minimum iron in diets. Proposed fines remained too low to deter malpractice, and relying on voluntary codes promised to delay welfare

[179] TNA MAF 369/75 Ruth Harrison Advisory Group, "Comment on the Minister's Proposals for Legislation following on the Brambell Report" (08.11.1966), enclosed in: Minute G.O. Lace to Mr Hensley (16.11.1966), 3-4.

[180] TNA MAF 369/77 "The Welfare of Farm Animals" (29 April), *The Times* (11.05.1967), 13A.

[181] TNA MAF 369/77 "The Welfare of Farm Animals" (29 April), *The Times* (11.05.1967), 13A.

improvements on farms.[182] However, political momentum for more ambitious reforms had ebbed. Announced in early November 1967 and passed in 1968,[183] the Agriculture (Miscellaneous Provisions) Bill complemented but did not replace the 1911 Cruelty to Animals Act by making it an offence to cause "unnecessary pain and distress" to livestock.[184] It also empowered MAFF to establish voluntary codes of practice and make statutory welfare regulations with regard to housing, feeding, and mutilation after consulting relevant interests.[185] The Act granted official veterinarians the right to enter and inspect farms.[186] However, there was no mention of mandatory space and nutritional requirements, and lighting standards, nor a concrete definition of what unnecessary pain and distress meant.[187] Decisions on controversial practices like the docking of pigs' tails were referred for further consultation to the new FAWAC.

Already established in late 1967, MAFF's new expert welfare body was composed of officials, scientists, industry representatives, and welfare campaigners—including RSPCA member Irene Walsh and Ruth Harrison.[188] Although the much-weakened 1968 Bill did not meet the demands of either *Animal Machines* or the 1965 Brambell Report, her nomination to FAWAC was a major personal success for Harrison. Previously belittled as an emotional housewife, she had skilfully used the political momentum of

[182] RSPCA Archives, Meeting of the Council, 23.11.1967, Attached: R.F. Seager to all members of the Council, 14.12.1967 and attached minutes of General Purposes committee, 02.11.1967 and enclosed letter: R.F. Seager to H.B. Fawcett (MAFF), 12.09. 1967.

[183] TNA MAF 369/77 Press Release MAFF (02.11.1967—1).

[184] *Agriculture (Miscellaneous Provisions) Act 1968*, Legislation.gov.uk; http://www.legislation.gov.uk/ukpga/1968/34 [09.01.2015].

[185] TNA MAF 369/90 Agriculture (Miscellaneous Provisions) Bill. Explanatory Memorandum. Part 1 Welfare Livestock.

[186] TNA MAF 369/90 Agriculture (Miscellaneous Provisions) Bill. Explanatory Memorandum. Part 1 Welfare Livestock; TNA MAF 369/272 Annex A—General Background, enclosed in: Storey to Mr Hann (13.03.1981), 1.

[187] TNA MAF 369/272 Annex A—General Background, enclosed in: Storey to Mr Hann (13.03.1981),1-3.

[188] TNA MAF 369/77 H.B. Fawcett to [anonymous] (18.09.1967); RSPCA Council had not been consulted about the appointment of Walsh, who had been a civil servant and served in a branch of the Secret Service during the Second World War. It subsequently tried to influence Walsh's actions and also tried to influence fellow FAWAC member Major Graham, vice-chairman of the Country Landowners' Association; RSPCA Archives, CM/57: 1966-1968; Meeting of the Council, 27.07.1968, 4; Meeting of the Council, 26.10.1967, 2-3.

her bestseller and her public status as a charismatic outsider to gain access to the confidential 'backstage' of British advisory committees. Remarkably, Harrison had done so without tethering herself to any of the major British agricultural, consumer, or welfare organisations. By stressing her position as an independent 'moderate' within the budding welfare movement, Harrison gave MAFF officials the impression of being an 'easy' choice to generate public acceptance for FAWAC and its upcoming welfare assessments. This impression soon proved incorrect.

PART IV

Defining Welfare (1967–1979)

Part IV analyses the increasingly crowded arena of animal welfare politics, activism, and science in post-*Animal Machines* Britain. The period between 1967 and FAWAC's dissolution in 1979 saw Harrison stubbornly push for the implementation of Brambell recommendations. This was no easy task. With animal-related politics becoming increasingly polarised, Harrison was attacked by radical activists for being too moderate and was side-lined by moderate campaigners and officials for being too radical. A fallout with the Royal Society for the Prevention of Cruelty to Animals (RSPCA) Council over hunting in 1970 resulted in a libel suit that forced Harrison to file for personal bankruptcy in 1975. All the while, British farm animal welfare politics were evolving rapidly. At the official level, FAWAC members and senior campaigners like Harrison worked within the framework of the 1968 Agricultural (Miscellaneous Provisions) Act to develop new welfare codes. This attempt to improve rather than abolish intensive practices was criticised by a younger generation of activists, who also challenged traditionalist leadership structures within the RSPCA. Sustained public controversy about animal welfare and rights triggered an expansion and professionalisation of campaigning. It also created an incubation chamber for a new "mandated"[1] form of animal welfare science tasked with evaluating and proposing welfare standards. The growing number of veterinary and behavioural researchers joining this field found that maintaining scientific credibility was difficult. Caught between competing conceptions of animal welfare, researchers struggled to satisfy hopes for universal welfare standards and navigate relations with non-governmental funders like the RSPCA and Ruth Harrison, who were intent on generating useful findings for their campaigning.

[1] David Fraser, "Understanding Animal Welfare," S1.

CHAPTER 8

A "minority of one": Harrison and the FAWAC

Installed in 1967, the Farm Animal Welfare Advisory Committee (FAWAC) was supposed to provide authoritative welfare advice. Following corporatist principles, MAFF had staffed the committee with a mix of veterinary and agricultural experts as well as nominees representing producer and welfare interests. Although FAWAC was supposed to produce compromise solutions, its overall membership was strategically weighted to favour agricultural interests and—despite initial plans—included no ethologist. As MAFF discussions about Harrison's nomination show (Chap. 7), the female welfare representatives had been chosen in the hope that they would not prove 'cranky' and generate public acceptance for FAWAC recommendations—neither hope was fulfilled.

Headed by Humphrey Robert Hewer, Professor of Zoology at Imperial College, FAWAC's first task was to prepare welfare codes to accompany the 1968 Agriculture (Miscellaneous Provisions) Bill. Although the codes were voluntary, they could be used to establish guilt in cruelty prosecutions brought against individual offenders. MAFF was also willing to consider statutory regulations governing the iron content of calf feeds and lighting provisions in sties as well as further regulations banning the bleeding of calves, the docking of cattle, and the docking of over three-day-old pigs without an anaesthetic.[1]

[1]TNA MAF 369/272 Annex A—General Background, enclosed in: Storey to Mr Hann (13.03.1981), 3.

C. Kirchhelle, *Bearing Witness*, Palgrave Studies in the History of Social Movements,
https://doi.org/10.1007/978-3-030-62792-8_8

Deciding on what to recommend proved divisive. From the beginning, FAWAC meetings were characterised by ideological clashes between agricultural and welfare representatives. Farming members blocked the adoption of recommendations from the Brambell Report, and welfarists criticised agricultural code proposals as insufficient.[2] Both sides refused to give way. Although FAWAC submitted code proposals to MAFF in September 1968,[3] the internal deadlock meant that many proposals remained vague. Reflecting the relative insignificance of domestic veal production, FAWAC agreed that "some of the husbandry methods involved in the production of 'white' veal are unacceptable on welfare grounds."[4] However, the committee issued no comments on other controversial aspects of intensive production such as antibiotic use or 'vice' amongst pigs and poultry.[5] Instead, FAWAC called for more research. Writing to Labour's Minister of Agriculture Cledwyn Hughes, FAWAC chairman Humphrey Hewer warned:

> The Codes are dealing with issues of a highly controversial nature; and in a few instances the Committee did not feel able to go all the way with the findings of the Brambell Committee because of the practical consequences … Some members, therefore, while agreeing that the Codes could be submitted to you for your agreement to circulation, reserve the right to raise these particular Brambell recommendations again when the comments of external organisations and individuals are available and the Committee are in a better position to assess the likely consequences of implementation of the various recommendations in the Codes.[6]

[2] TNA MAF 369/272 Annex A—General Background, enclosed in: Storey to Mr Hann (13.03.1981), 4.

[3] TNA MAF 369/272 Annex C—H.R. Hewer to Minister Cledwyn Hughes (06.09.1968), enclosed in: Storey to Mr Hann (13.03.1981).

[4] TNA MAF 369/272 Annex C—H.R. Hewer to Minister Cledwyn Hughes (06.09.1968), enclosed in: Storey to Mr Hann (13.03.1981), 1; the proposed ban of intensive veal production was ultimately blocked by MAFF due to legal concerns about having to stop imports during a time when Britain was trying to join the EEC; TNA MAF 369/272 Annex B, enclosed in: Annex C—H.R. Hewer to Minister Cledwyn Hughes (06.09.1968), enclosed in: Storey to Mr Hann (13.03.1981).

[5] TNA MAF 369/272 Annex C—H.R. Hewer to Minister Cledwyn Hughes (06.09.1968), enclosed in: Storey to Mr Hann (13.03.1981), 2–3.

[6] TNA MAF 369/272 Annex C—H.R. Hewer to Minister Cledwyn Hughes (06.09.1968), enclosed in: Storey to Mr Hann (13.03.1981), 2.

MAFF subsequently circulated FAWAC's code proposals to about 140 organisations for comment before submitting a weakened version of the initial proposals to Parliament in October 1969.[7]

Welfare campaigners were furious about the perceived weakening of key Brambell recommendations. In an extremely damaging move for FAWAC, four committee members publicly attacked the proposed codes. The dissenters were Joan Maynard from the National Union of Agricultural Workers, Dorothy Sidley from the Humane Slaughter Association, the RSPCA's Irene Walsh, and Ruth Harrison. The four female campaigners expressed their "disappointment that some practices and systems have been considered acceptable which … we recognize must inevitably cause prolonged discomfort and probably mental suffering."[8] Justifying her decision not to resign from FAWAC in protest, Harrison noted, "I think the committee has made a genuine effort to improve the welfare of animals, but they have not gone far enough."[9]

Former Brambell Committee members, including Rogers Brambell and W.H. Thorpe, also attacked the new codes. In June 1969, they criticised the codes for failing to adhere to central Brambell recommendations. The codes did not guarantee sufficient freedom of movement for an animal to turn around, groom itself, lie down, and stretch; ensure the provision of sufficient food, water, and flooring for an animal to feel secure on; or mandate suitable ventilation, inspection, and environmental emergency measures. Proposed stocking densities for fowl and turkeys were particularly shocking: "These densities defined in the Codes are a compromise on a compromise for which no case otherwise than commercial expediency exists."[10]

A follow-up letter to the *Times* by leading animal researchers was even more scathing. Signed by Julian Huxley, Jon R. Baker, James Fisher, Alister Hardy, Desmond Morris, Niko Tinbergen, J.W.S. Pringle, Peter Scott, O.L. Zangwill, and Laurence Weiskrantz, the letter doubted that the FAWAC's codes could protect animals' physical and emotional welfare: "As scientists familiar with the behaviour of animals we feel strongly, with the Brambell Committee, that every possible step must be taken to

[7] TNA MAF 369/164 Animal Health Division II, FAWAC, Meeting with Professor Hewer on 14.08.1970, Note for Minister (05.08.1970), 1.

[8] Leonard Amey, "Farming Notes. Factory farming code contention", *Times*, 28.07.1969, 10.

[9] J.W. Murray, "Woman writer protests over factory farming", *Observer*, 29.06.1969, 3.

[10] F.W. Rogers Brambell et al. "Codes For Factory Farming", *Times*, 23.06.1969, 9.

prevent" confinement that frustrates "major activities which make up its natural behaviour."[11] Doing so would protect British animals and morals:

> As citizens of a modern nation we are further convinced that the practice of keeping animals under severely frustrating conditions, with all the signs of incipient or full nervous disfunction similar to those of distress in ourselves, must have a numbing effect on the farmer's own sensitivity. (…) children who grow up in a society in which distress in animals is recognized, yet tolerated or ignored, may well develop a generally callous attitude in later life. (…). In the long term the tolerance, by a civilized society, or cruelty to animals which is recognized as such seems to us to carry the danger of returning to the level of the barbarian to whom animals are things rather than fellow creatures.[12]

Remarkably, the letter had not been drafted by Julian Huxley, who publicly appeared as its main author, but by Niko Tinbergen, who seemed to be abandoning his earlier refusal to publicly engage with animals' affective states.[13]

Faced with this barrage of criticism,[14] FAWAC chairman Humphrey Hewer responded with his own letter to the *Times*. A close reading of the codes showed that they were not cruel but followed most Brambell recommendations—calves could turn around, docking was only permitted by surgeons following injury and disease, and the provision of bedding was recommended. Since Parliament had failed to pass mandatory welfare recommendations, the new voluntary codes were the best way forward and could be used to aid prosecutions: "I would have thought that animal welfarists would have welcomed the Codes and not obstructed them."[15]

Ahead of the first House of Lords debate on the FAWAC codes, Ruth Harrison increased public pressure for more stringent recommendations with an article in the *Observer*. Titled "Why Animals Need Freedom to Move,"[16] the full-page article portrayed conditions on factory farms:

[11] Julian Huxley et al, "Factory Farming", *Times*, 25.06.1969, 11.

[12] Julian Huxley et al, "Factory Farming", *Times*, 25.06.1969, 11.

[13] Rice University, Julian Sorrell Huxley Papers, Julian Huxley to Nikolaas Tinbergen, 08.05.1969, Handwritten response Tinbergen.

[14] See also F.W. Rogers Brambell, "Codes for Factory Farming Rules Without Force", *Times*, 01.07.1969, 9; R.F. Seager [RSPCA], "Chance to act", *Times*, 01.07.1969, 9.

[15] H.R. Hewer, "Force in Codes for Factory Farming", *Times*, 05.07.1969, 9.

[16] Ruth Harrison, "Why Animals Need Freedom To Move", *Observer*, 12.10.1969, 7.

> Lay this copy of The Observer on the floor. That is the living space of five fully grown hens ... Now open out two pages and lay them end to end. That is the living space (5ft by 2 ft) of a 'white veal' calf. Add another half page and you have the space allowed to a pregnant sow (2ft by 7ft).[17]

Harrison also used a Gallup poll she had recently commissioned to insist that "the majority of farmers, the overwhelming mass of public opinion, ... is urging that animals be allowed freedom of movement."[18] According to Harrison, "the passage in 'Animal Machines' which angered agricultural spokesmen most was the one claiming that cruelty is acknowledged only when profitability ceases. Yet in essence this remains as true today as it was then."[19] Although it was unrealistic to expect a complete return to the extensive systems praised in her book, it was necessary to "take an ethical stand that any codes must be amended so as to be at least equal to the minimum standards laid down by the Brambell Report."[20] Government and consumers needed to encourage farmers to break the cycle of further intensification.

Public protest against the proposed welfare codes was successful. During stormy parliamentary debates, speakers accused FAWAC and MAFF of making inadequate provisions for flooring, space allowances, bedding, freedom of movement, surgical operations, and feeding.[21] An internal MAFF summary of resulting press coverage acknowledged, "a lot of critics made no bones about their view that [FAWAC] was composed of vested interests determined to see that welfare considerations did not interfere with economic aims."[22] Under significant pressure in view of welfare protests (Chap. 9) and an upcoming critical report on agricultural antibiotic use,[23] Minister Hughes spontaneously undertook to resubmit

[17] Harrison, "Why Animals Need Freedom To Move".

[18] Harrison, "Why Animals Need Freedom To Move".

[19] Harrison, "Why Animals Need Freedom To Move".

[20] Harrison, "Why Animals Need Freedom To Move".

[21] TNA MAF 369/164 Press Notice MAFF, Animal Welfare—Welfare of Livestock, Notes for Editors (29.09.1970), 2; Woods, "Cruelty to Welfare," 19–21.

[22] TNA MAF 369/164 Animal Health Division—Meeting with Prof Hewer on 14.08.1970. Note for Minister (05.08.1970), 2.

[23] Claas Kirchhelle, "Swann song: antibiotic regulation in British livestock production (1953–2006)," *Bulletin of the History of Medicine* 92/2 (2018), 317–350.

contested elements of the codes to FAWAC[24] and commissioned a welfare survey by the State Veterinary Service (SVS).[25]

Both organisations reported back in 1970. Based on 4690 field visits revealing 36 cases of "unnecessary" suffering and pain, the SVS report was published in September 1970 and claimed that "the standard of stockmanship on intensive units is sound."[26] FAWAC's code review was altogether less smooth. Following the dissenters' media campaign, two members had resigned in September 1969, and a further member did not wish to renew membership.[27] Wary of the hostile mood within FAWAC, Prof. Hewer warned MAFF that "the full Committee would find it difficult to draw up a view on the disputed points."[28] In order to diffuse tensions and produce some kind of review, Hewer "proposed that two "drafting groups" should be set up to expand the respective opinions of the groups holding divergent views."[29] The committee was thus split into two review groups. Reflecting an agro-industrial bias of Hewer and MAFF officials, the two FAWAC groups were initially referred to as 'majority' and 'minority' groups and later renamed 'scientific' and 'ethical' groups.[30]

Amongst the practices to be reviewed by the two groups were withholding roughage, space standards for singly penned cattle, prolonged or continuous tethering, bedding requirements, and slatted floors for cattle. Another area of inquiry focused on sow stalls, the provision of bedding, and slatted floors for pigs. Reviewers also queried space allowances, spectacles, beak trimming, dubbing, and skip-a-day feeding for poultry. The two review groups would also assess the freedom to turn round and dim lighting for all animals.[31]

[24] TNA MAF 369/272 Annex A—General Background, enclosed in: Storey to Mr Hann (13.03.1981), 5.

[25] TNA MAF 369/164 Press Notice MAFF, Animal Welfare—Welfare of Livestock, Notes for Editors (29.09.1970), 2; TNA MAF 369/163/2 Cledwyn Hughes to Prof Hewer [04.12.1969].

[26] TNA MAF 369/164 Press Notice MAFF, Animal Welfare—Welfare of Livestock, Notes for Editors (29.09.1970), 1.

[27] TNA MAF 369/164 Animal Health Division Meeting with Professor Hewer on 14.08.1970—Note for Minister (05.08.1970), Appendix A.

[28] TNA MAF 369/163/2 Note of meeting with Prof Hewer (30.10.1969).

[29] TNA MAF 369/163/2 Note of meeting with Prof Hewer (30.10.1969).

[30] TNA MAF 369/164 Letter from H.R. Hewer (21.08.1970); draft enclosed in: H.B. Fawcett to Ruth Harrison (30.11.1970).

[31] TNA MAF 369/163/2 Cledwyn Hughes to Prof Hewer [04.12.1969].

At stake was not only the viability of FAWAC as an official body but also the relative weight that behavioural welfare indicators should have in its decision-making. Conflicts about the status of ethological expertise characterised both reviews from the beginning. Similar to later badger culling controversies, MAFF officials tended to discount research on 'feelings' and behaviour as 'soft' and anthropomorphic. Instead, they favoured traditional agricultural and veterinary expertise and 'hard' physiological evidence. The "epistemic rivalry"[32] between the animal health and ethological communities was made evident in FAWAC's linguistic conflation of 'ethics' and 'ethology,' which were both considered beyond the purview of the 'scientific/majority' review.[33]

Together with the RSPCA's Mid-Wales secretary Irene Walsh, Ruth Harrison headed the four-person all-female 'minority' or 'ethical' group. Meeting for the first time on January 13, 1970, members regretted the majority group's tendency to dismiss ethical and behavioural considerations and the biased way in which unpublished evidence had been used to draft the initial code proposals:

> Ethical considerations and observational evidence by laymen were important and should be taken fully into account. The Group felt that the Advisory Committee, while not giving sufficient weight to the ethical approach, had nevertheless not been consistently scientific in its attitude. Some of its conclusions had been based on economic considerations rather than on science.[34]

The ethical group specifically bemoaned the absence of an ethologist on FAWAC. This lack of ethological input and the industry bias of the majority group had been particularly glaring during a meeting with rumination experts in February 1969:

> Originally the Advisory Committee intended to discourage white veal production. This was subsequently reversed largely on the basis of the opinions expressed by the visiting scientists. … Little evidence was available on the

[32] Cassidy, *Vermin*, 227; for a fuller discussion of what Cassidy terms cultures of care see, 75–102; Robert Kirk has highlighted a similar neglect of ethological expertise by the UK's Chief Scientific Adviser Solly Zuckerman in the case of laboratory animals, Kirk, "Clinic and Laboratory," 525–526.

[33] I am indebted to Henry Buller for this observation.

[34] TNA MAF 369/163/2 Draft—FAWAC—Re-Examination of Disputed Points in Codes. Minority Drafting Group. Record of the Group's first meeting (13.01.1970), 1–2.

> highly important behavioural implications of roughage denial. These had been largely ignored in reaching the final decision.[35]

Similar problems had affected the drafting of poultry codes, which had legitimised dim lighting as a dubious corrective to behavioural problems and given the impression that "management is not a critical factor at all densities."[36]

The majority group disagreed. Including C. Graham of the Country Landowners' Association, physiologist Morrell Draper of the Agricultural Research Council's Poultry Research Centre, and veterinary researcher David Sainsbury, members of the group were sceptical of 'soft' behavioural welfare indicators. Instead, they favoured 'hard' physiological measures like protein metabolism as "the best available indicator of well-being."[37] A refined version of the thrift argument, the protein metabolism theory assumed that inadequate welfare would disrupt protein synthesis and growth. It was thought that protein synthesis was a more accurate indicator of welfare than weight gain, which could also be caused by the accumulation of water and fat. Attacking the minority group's ethical arguments, the majority group claimed:

> The problem [is] that ethical arguments introduced subjective considerations of how far the benefit of any doubt should be given to the animal and how far economic factors should be allowed to limit the translation of ethical principles into practice. The strictly scientific approach avoided or minimized these difficulties.[38]

According to this view, there was no conclusive evidence necessitating an amendment of welfare codes for tethering, the provision of bedding, keeping animals on slats, or beak trimming to prevent vice. Discounting the ethological argumentation underpinning the Brambell Report, "[David] Sainsbury told of calves reared in Holland which had become so

[35] TNA MAF 369/163/2 Draft—FAWAC—Re-Examination of Disputed Points in Codes. Minority Drafting Group. Record of the Group's first meeting (13.01.1970), 3.

[36] TNA MAF 369/163/2 Draft—FAWAC—Re-Examination of Disputed Points in Codes. Minority Drafting Group. Record of the Group's first meeting (13.01.1970), 5.

[37] TNA MAF 369/163/2 FAWAC. Majority Group. Notes of first meeting to consider disputed points (04.02.1970), 1.

[38] TNA MAF 369/163/2 FAWAC. Majority Group. Notes of first meeting to consider disputed points (04.02.1970), 3.

accustomed to a small pen that, even when given extra space, had not turned around."[39] The group also rejected updating space requirements for battery hens and cattle.[40]

Attempting to refute the protein metabolism argument, Ruth Harrison made various information requests through FAWAC's secretariat. If the majority group was unwilling to accept ethical arguments, she would need further behavioural and physiological data to support an increasingly concrete concept of good welfare as a positive state encompassing more than the absence of pain and stress.[41] Most of Harrison's initial information requests centred on determining haemoglobin levels in veal calves and traditionally reared calves to ascertain whether intensive veal production induced anaemia. Writing to FAWAC secretary H.B. Fawcett in February 1970, she complained about the lack of detail in expert reports, which had been sent to FAWAC members: "I do my best to worship at the altar of scientific evidence, but I must say that at times my faith is sorely tried!"[42] In his response, the supposedly neutral Fawcett doubted Harrison's ability to interpret scientific evidence:

> As regards worshipping at the altar of scientific evidence it is sometimes salutary to remember Bertrand Russell's advice in his 'essay on scepticism'. In this he suggested that the common man (that is you and me in this context) would be prudent if, when experts agreed, he did not hold the contrary opinion, and when experts disagreed he suspended his judgement.[43]

Fawcett's response failed to deter Harrison. In the following months and years, she was unique amongst FAWAC members regarding her high volume of information requests.[44] Unsurprisingly, this behaviour did not

[39] TNA MAF 369/163/2 FAWAC. Majority Group. Notes of first meeting to consider disputed points (04.02.1970), 7.

[40] TNA MAF 369/163/2 FAWAC. Majority Group, Notes of second meeting to consider disputed points (23.02.1970), 2.

[41] TNA 369/163/2 H.B. Fawcett to Ruth Harrison (19.01.1970); Broom, "Ruth Harrison's later writings and animal welfare work".

[42] TNA 369/163/2 Ruth Harrison to H.B. Fawcett (04.02.1970), enclosed in: H.B. Fawcett to Ruth Harrison (05.02.1970).

[43] TNA MAF 369/163/2 H.B. Fawcett to Ruth Harrison (05.02.1970).

[44] TNA MAF 369/163/1 D.J. Kotulanski to Miss B.F. Moore (21.05.1970); Minute [handwritten] Kotulanski to Mr Goaten (01.04.1970); Minute [handwritten] Mr Foreman (21.05.1970).

endear her with MAFF officials. Following further information requests by Harrison, a MAFF memo stated:

> [Harrison] was 'tactful enough' [sic] to say that if her requests resulted in too much work or inconvenience not to hesitate to say no. Tactfully, I hope, I thanked her for encouraging us to say 'no' to her requests, but pointed out that normally we always try to assist anybody if we can do so.[45]

Harrison's steadfast refusal to accept (male) expert opinion without supporting data was impressive. However, her single-mindedness could also have a more problematic side. Her relentless pursuit of the goals set out in *Animal Machines* also meant that the self-characterised "loner"[46] was at times willing to cut corners and alienate allies if agendas diverged. In the case of FAWAC, Harrison began requesting information and trips without informing fellow minority group members. This behaviour did not go unnoticed by MAFF officials. When Harrison requested access to communications between experts and the majority group and asked for individual slaughterhouse tours, an official asked her "whether she was making her request on behalf of the Minority Group members—she hesitated and said that she understood Mrs Walsh had intended to make a similar request, otherwise she was speaking for herself."[47] Another minute warned, "I do not think that we should become involved in helping one particular member without the knowledge of the group itself."[48]

While her tendency to act as a "minority of one"[49] cost Harrison support from fellow animal campaigners (Chap. 9), it was remarkably effective in the corporatist context of Whitehall politics. The 1969 media campaign against FAWAC's code proposals had taught Harrison that thwarting weak consensus reports and instigating public pressure could be a useful way of countering pro-industry biases in official circles. The following years would see her employ this strategy again and again.

In the context of the 1970 FAWAC welfare code review, Harrison clashed with majority group members on issues ranging from the tethering of sows to the composition of animal feeds. Mentioned more than any other participant in FAWAC's minutes, she systematically vetoed potential

[45] TNA MAF 369/163/1 Minute [illegible, handwritten] to Mr Foreman (15.05.1970), 2.
[46] Kendall, "Ruth and the Ruthless," 21.
[47] TNA MAF 369/163/1 Minute [handwritten] to Mr Foreman (15.05.1970), 1.
[48] TNA MAF 369/163/1 Minute Mr Foreman to Mr Fawcett (15.05.1970).
[49] Oral History Interview Marian Stamp Dawkins (01.07.2014).

compromises between majority and minority views. When one of her main opponents, the dairy farmer W.A. Bigger, expressed opposition to the use of terms that "gave a false impression that members' views were worlds apart" in April 1970, Ruth Harrison insisted "that there were satisfactory alternative methods and some producers were turning to them."[50] Referring to Harrison, physiologist Morrell Draper warned "that the Minority Group's reluctance to accept well-founded scientific opinion augured little hope of its acceptance of any opinion differing from its own."[51]

Ruth Harrison's unwillingness to yield ground led to the publication of three separate FAWAC review documents: a majority report, a minority report, and a separate comment by Ruth Harrison.[52] The major point of contention remained whether animals' behaviour and affective states deserved consideration. In their final report, the minority group opposed both the original FAWAC codes and protein metabolism as an absolute indicator of animal welfare:

> We believe that it is essential to take ethical considerations fully into account; (…) the present codes set bare minimum standards. Indeed, in many cases they fail to make even a first step towards achieving the desired welfare objective.[53]

Rejecting the "Brambell view that animals should be given the benefit of the doubt,"[54] the majority group countered that ethics and subjective feelings had no role in scientific standard-setting:

[50] TNA MAF 369/80 Draft, Minutes of the Seventh Meeting of the Committee (22.04.1970), 15.

[51] TNA MAF 369/80 Draft, Minutes of the Seventh Meeting of the Committee (22.04.1970), 18.

[52] TNA MAF 369/163/1 FAWAC, Re-Examination of Disputed Recommendations in Welfare Codes. To Be Read in Conjunction with Report by Minority Group—Author Ruth Harrison.

[53] TNA MAF 369/163/1 FAWAC Re-Examination of Disputed Recommendations in Welfare Codes, Office Note, Report by Minority Group, 1.

[54] TNA MAF 369/163/1 FAWAC Re-Examination of Disputed Recommendations in Welfare Codes, Office Note, Report by Majority Group, 14.

> By avoiding subjective argument and the dangers of the anthropomorphic approach, [the scientific approach] provides the firmest available basis for reaching decisions on welfare issues.[55]

In her separate comment, Ruth Harrison attacked FAWAC's welfare record. According to Harrison, the majority group's protein metabolism argument was discredited by the simple fact that technologies like antibiotics enabled animal growth even in adverse circumstances. By focusing only on acute pain and suffering, FAWAC was defining welfare as the absence of cruelty instead of setting out a positive vision of physical and mental well-being. "At every point," FAWAC had legally "given the producer rather than the animal the benefit of the doubt":

> Although the codes are purely voluntary we have drafted them with the care of mandatory regulations. We have been afraid of excluding any single system or technique if it can be managed successfully by any one producer … What we have also tended to overlook is that just as non-compliance can tend to establish guilt, compliance can help to establish innocence, the standards set should be high enough to allow the definition in the Act to have some clear meaning.[56]

Referring to the majority group's emphasis on scientific evidence, Harrison reminded members that both the Brambell Committee and FAWAC had been established because of the British public's ethical concerns about animal welfare:

> the fact that an animal has limbs should give it the right to use them, the fact that a bird has wings should give it the right to spread them, the fact that both are mobile should give them the right to turn round and the fact that they have eyes should give them the right to see.[57]

The ongoing stalemate between FAWAC's 'ethical' and 'scientific' factions meant that no consensus was reached regarding the prolonged tethering of cattle, slatted floors, bedding for cattle and pigs, sow stalls,

[55] TNA MAF 369/163/1 FAWAC Re-Examination of Disputed Recommendations in Welfare Codes, Office Note, Report by Majority Group, 2–3.

[56] TNA MAF 369/163/1 FAWAC, Re-Examination of Disputed Recommendations in Welfare Codes. To Be Read in Conjunction with Report by Minority Group—Author Ruth Harrison, 3.

[57] TNA MAF 369/163/1 FAWAC, Re-Examination of Disputed Recommendations in Welfare Codes. To Be Read in Conjunction with Report by Minority Group—Author Ruth Harrison, 4.

space allowances for poultry, roughage for calves, and lighting in sties.[58] The only points both sides agreed upon were (1) that poultry blinkers and spectacles constituted a mutilation of the nasal septum and should be banned; (2) that the width of cattle pens should be equal to animals' height so that animals could groom themselves, lie down, and fully extend their limbs; (3) that skip-a-day feeding systems were acceptable.[59]

Fearing public reactions to another failure to update codes, many FAWAC members were reluctant to publish their deliberations.[60] Writing to Labour Minister of Agriculture Cledwyn Hughes ahead of the 1970 General Election, FAWAC Chairman Hewer reported that members feared it would be "difficult to ignore the public criticism (pressure would be a more apt term), to which [FAWAC] has already been subjected to some extent."[61] Hewer recalled a recent FAWAC meeting with "a group of MP's led by [Conservative MP and RSPCA Council member Frederick] Burden":

> In the event, although Mr. Burden and his colleagues provided no evidential support for their welfare opinions, I believe that the meeting served one useful purpose in that the MPs cannot claim that they have had no opportunity to express their views to us directly. I think, too, that my members, by exercising great restraint and reasonableness in the face of a good deal of provocation (sometimes bordering on rudeness), enhanced the status of the Committee.[62]

However, Hewer's protest was of no avail. Following pressure from the newly elected Conservative Minister of Agriculture, James Prior, the three FAWAC reports were published alongside the SVS report in late 1970.[63]

Her success in blocking a compromise on welfare code revisions turned into a political victory for Ruth Harrison. Although Parliament eventually passed modified versions of the original codes (see Chap. 9), MPs' criticism of the contradictory 'majority' and 'minority' reports led to an

[58] TNA MAF 369/163/1 FAWAC Re-Examination of Disputed Recommendations in Welfare Codes, Office Note, Report by Minority Group, 10.

[59] TNA MAF 369/163/1 Draft, Re-Examination of Disputed Recommendations in Welfare Codes. Points Upon Which the Whole Committee Is Agreed.

[60] TNA MAF 369/163/1 H.R. Hewer to Rt. Hon. Cledwyn Hughes (04.05.1970), 1.

[61] TNA MAF 369/163/1 H.R. Hewer to Rt. Hon. Cledwyn Hughes (04.05.1970), 2.

[62] TNA MAF 369/163/1 H.R. Hewer to Rt. Hon. Cledwyn Hughes (04.05.1970), 2.

[63] TNA MAF 369/164 H.B. Fawcett to Ruth Harrison (30.11.1970); Letter from H.R. Hewer (21.08.1970); Minute A.C. Sparks to Mr Murphy (14.08.1970); Woods, "Cruelty to Welfare," 20.

extension of FAWAC's terms of reference to include all animals kept on agricultural land. FAWAC also began debating whether to recommend regulations banning the docking of cattle and the dewinging and castration of poultry.[64] Resulting ban recommendations were partially enacted as enforceable regulations in 1974.[65] Within FAWAC, Harrison also pushed for a greater role for behavioural expertise and managed to secure invitations for renowned researchers like Nikolaas Tinbergen and Ingvar Ekesbo[66] as well as a new working group on wing injuries in battery cages.[67]

However, preventing FAWAC consensus on weak welfare codes was not the same as finding robust majorities for a sustained strengthening of codes. Despite Harrison's efforts, the overall pace of FAWAC deliberations and of British welfare reforms remained glacial throughout the 1970s. In the rare cases that FAWAC members jointly expressed concern about new husbandry systems, it did not necessarily lead to MAFF preventing their introduction.[68] Meagre resources further hampered FAWAC's work. Although its research sub-committee had been tasked with identifying animal welfare research priorities, FAWAC was unable to provide funds for any of the 30 resulting research questions.[69]

MAFF was the main profiteer of FAWAC's impasse. When pressed on welfare issues, Ministers referenced the uncertain "present state of scientific knowledge"[70] to fend off calls for statutory regulations. MAFF's lack of engagement with welfare issues extended to its enforcement of already enacted codes. During the mid-1970s, it emerged that the limited existing welfare codes were not being observed on some farms and not adequately

[64] TNA MAF 369/80 (180) Draft, Minutes of the Eighth Meeting of the Committee (10.02.1971), 3–6; Woods, "Cruelty to Welfare," 21.

[65] TNA MAF 369/204 Draft Statutory Instruments 1974, Animal Prevention of Cruelty. The Welfare of Livestock (Cattle and Poultry) Regulations 1974; TNA MAF 369/272 House of Commons Agriculture Committee—Replies to Questions Enclosed With Dr Jack's Letter of 03.02.1981 to Mr Shillito, 4.

[66] TNA MAF 369/80 (180) Draft, Minutes of the Eighth Meeting of the Committee (10.02.1971), 10; TNA MAF 369/206 Annex B to AWC/74/Mins 2.

[67] TNA MAF 369/80 (180) Draft, Minutes of the Eighth Meeting of the Committee (10.02.1971), 5 and 12–13.

[68] TNA MAF 369/204 MAFF, Welfare of Livestock. Animal Welfare. Early Weaning and Cage Rearing of Piglets (24.08.1973).

[69] TNA MAF 369/204 FAWAC Minutes of 10th Meeting (22.11.1972), 10.

[70] TNA MAF 369/204 MAFF—Priority Written Question No. 25 (20.06.1974); Woods, "Cruelty to Welfare," 21.

controlled by officials. Understaffed by ca. 20 per cent,[71] the SVS was failing to perform up to 75 per cent of its welfare duties.[72] In 1978, MAFF disclosed that between 1968 and 1977 it had initiated only 11 welfare prosecutions under the 1968 Bill.[73]

While British welfare reform stagnated, other European states enacted ambitious new regulations. In 1972, West Germany passed a Protection of Animals Act, which mandated that any person keeping or supervising animals should give them adequate food and care suitable for their species. Explicitly referencing new concepts of welfare, the act also mandated that accommodation was to be provided according to an animal's natural behaviour.[74] In Sweden, all animals had to be effectively stunned prior to slaughter, and strict rules governed the transport of live and pregnant animals since the early 1970s.[75] In Norway, the 1974 Animal Protection Act stated that animal's "instincts and natural needs" should be taken into consideration to avoid unnecessary suffering.[76]

Growing European market integration exerted pressure on Britain to accept common standards and join transnational welfare bodies. In November 1972—about one month ahead of Britain's accession to the EEC—FAWAC was informed of a proposed Council of Europe Convention on Animal Welfare in Intensive Rearing, which had been debated at a meeting in Strasbourg.[77] The proposed convention would apply welfare standards based on physiological and behavioural parameters to farm animals throughout Europe and create a standing group on welfare at the European level. As official representatives of the UK, MAFF's two FAWAC secretaries had been present at the Strasbourg meeting alongside Ruth Harrison, who had attended as an observer for the World Federation for

[71] TNA MAF 369/208 MAFF Press Notice—Parliamentary Secretary Speaks At BVA Congress (15.09.1975).

[72] TNA MAF 369/204 Ruth Harrison to Professor Hewer (13.01.1974); Minute G.B. Taylor to J.N. Jotcham (20.02.1974); TNA MAF 369/222 PQ 5032. MAFF—Parliamentary Question (13.12.1977).

[73] TNA MAF 369/240 PQ 5801. MAFF—Parliamentary Question (05.05.1978).

[74] FACT Files, MD, FACT Publications & Publicity Material, Ruth Harrison, 'Introduction', Proceedings of Workshop sponsored by FACT and the UFAW: Behavioural needs of Farm Animals, *Applied Animal Behaviour Science*, 19 (1988), 342.

[75] Ingvar Ekesbo, "The Swedish Approach," in Council Of Europe (ed.), *Animal Welfare* (Strasbourg: Council of Europe, 2006), 185–86.

[76] Kristin Asdal and Tone Druglitrø, "Modifying the Biopolitical Collective: The law as a moral technology," in Kristin Asdal, Tone Druglitrø, Steve Hinchliffe (eds.), *Humans, Animals and Biopolitics* (London: Routledge, 2016), 66–84.

[77] TNA MAF 369/204 FAWAC Minutes of 10th Meeting (22.11.1972), 9.

the Protection of Animals (WFPA). Reporting on the meeting, Ruth Harrison was

> agreeably surprised to note that most countries represented were prepared to take a flexible approach to [animal welfare]. She felt that we had much to learn from some European countries; Norway, for instance, had already introduced a series of regulations covering battery cages.[78]

FAWAC received a draft of the proposed European Convention for the Protection of Animals Kept for Farming Purposes in late 1974. Despite remaining vague, both the draft and the resulting 1976 Convention acknowledged the welfare principles set out in the 1965 Brambell Report: animals were to have freedom of movement appropriate to avoid "unnecessary suffering or damage."[79] A tethered animal should have space "appropriate to its physiological and ethological needs in accordance with established experience and scientific knowledge."[80] In indoor settings, the same definition of need would also govern the provision of lighting, temperature, humidity, air circulation, ventilation, and "other environmental conditions."[81] Answering long-standing welfare criticism, animals were to be provided with food or liquid in a manner that would not cause "unnecessary suffering or damage."[82] On farms, animals' health and technical equipment were to be inspected by farmers at "intervals sufficient to avoid unnecessary suffering."[83] On intensive farms this meant at least once a

[78] TNA MAF 369/204 FAWAC Minutes of 10th Meeting (22.11.1972), 9; on the further development of Norwegian reforms leading up to initial welfare upgrades in 1982 see Kristian Bjørkdahl, "When the Battery Cage Came to Norway: The Historical Path of an Agro-Industrial Artifact," in Kristian Bjørkdahl and Tone Druglitrø (eds), *Animal Housing and Human-Animal Relations* (London: Routledge, 2016), 55–78.

[79] TNA MAF 369/204 Appendix I—Draft European Convention for the Protection of Animals Kept for Farming Purposes (EXP/An(74)4; Annex to AWC/74/10), 2.

[80] TNA MAF 369/204 Appendix I—Draft European Convention for the Protection of Animals Kept for Farming Purposes (EXP/An(74)4; Annex to AWC/74/10), 2.

[81] TNA MAF 369/204 Appendix I—Draft European Convention for the Protection of Animals Kept for Farming Purposes (EXP/An(74)4; Annex to AWC/74/10), 2.

[82] TNA MAF 369/204 Appendix I—Draft European Convention for the Protection of Animals Kept for Farming Purposes (EXP/An(74)4; Annex to AWC/74/10), 3; the word "damage" was later replaced with injury; "European Convention for the Protection of Animals Kept for Farming Purposes (Strasbourg, 10.03.1976)", http://conventions.coe.int/Treaty/EN/Treaties/Html/087.htm [17.12.2014].

[83] European Convention for the Protection of Animals Kept for Farming Purposes (Strasbourg, 10.03.1976).

day.[84] Future animal welfare provisions were to be decided by a European standing committee on animal welfare.

While the Council of Europe convention marked an important formalisation of Brambell welfare principles, its standing committee posed a direct challenge to FAWAC authority. Ahead of Britain's ratification of the Council of Europe Convention,[85] MAFF considered it "inadvisable to open a direct channel of communication between the Standing Committee"[86] and FAWAC. FAWAC's role was downgraded to advise Ministers on policy lines for the Standing Committee, which met regularly between 1979 and 1998.[87]

In addition to being superseded at the European level, FAWAC also faced growing domestic pressure. In October 1973, the *New Scientist* printed a series of reports on animal welfare in Britain. In his contribution, Tom Ewer, former member of the Brambell Committee and professor of animal husbandry at the University of Bristol, launched a scathing attack on FAWAC. According to Ewer, FAWAC and MAFF inaction had led to the abandonment of large parts of the Brambell Report:

> We must all regret that the government, through FAWAC and the reports it has failed to write and publish, has dismally failed to reform and educate. The research has not been sponsored, nor the right questions asked, nor the relevant new information publicised. The advisory role that could be played by the state veterinarian, who has the right of entry, is grievously diminished because of acute undermanning; so even the great good our imperfect codes could achieve, is squandered.[88]

Although Britons were fond of highlighting their "sensitive national conscience"[89] regarding animals, European partners were enacting far more progressive welfare regulations. The proposed Council of Europe Convention offered some chance for improvement. Hoping that Britain would "play a leading part in forming the important standing committee,"

[84] This requirement was not covered by Britain's 1968 Agricultural (Miscellaneous Provisions) Act; TNA MAF 369/215 A. Foreman to FAWAC (03.05.1976).

[85] TNA MAF 369/215 A. Foreman to FAWAC (03.05.1976).

[86] TNA MAF 369/217 FAWAC, Minutes of 15th Meeting (22.06.1976), 6–7.

[87] TNA MAF 369/240 FAWAC, 18th Meeting (08.02.1979), 2.

[88] TNA MAF 369/204 Tom Ewer, "Farm animals in the law", *New Science* (18.10.1973), 179, enclosed in: Minute A. Foreman to FAWAC (08.11.1973).

[89] TNA MAF 369/204 Tom Ewer, "Farm animals in the law", *New Science* (18.10.1973), 179, enclosed in: Minute A. Foreman to FAWAC (08.11.1973).

Ewer nonetheless doubted that "FAWAC is the appropriate national 'advisory body'"[90] to provide relevant advice.

For Ruth Harrison, Britain's adoption of the Council of Europe Convention posed a dilemma. Although she endorsed both the Convention and the Standing Committee, the displacement of FAWAC threatened to undermine her own access to Whitehall policy circles. Having played no small part in paralysing FAWAC,[91] she tried to salvage its reputation by calling on the committee's ailing chairman to reinvigorate welfare reviews. Writing to Prof. Hewer in January 1974, she warned "that the Codes for which FAWAC is best known, are being widely disregarded and our position has become, on the face of it, embarrassing if not slightly ridiculous."[92] Britain was already "lagging far behind other European countries."[93] While it would be a "terrible pity if Britain has to take a back seat on the Council of Europe" Committee, it was even more worrying that the Minister of Agriculture did not seem to realise "how far we have fallen behind others."[94] In a somewhat bizarre statement, Harrison assured Hewer that she was actively defending FAWAC against British and European critics:

> I write in friendship, not in hostility, I have a good deal of faith in your knowledge and in the rest of the Committee. Please let us put all this to use in actually achieving a change on the farm. Let us be seen to be vigorous and active.[95]

Harrison's élan was not reciprocated. In February 1974, the terminally ill Hewer informed MAFF that there would be no new FAWAC measures prior to the appointment of a successor.[96] Referring to Hewer's announcement and Harrison's letter, a MAFF minute snidely remarked, "Mrs. Harrison's difficulty over [defending FAWAC] is probably no greater than our difficulty in attempting to deal rationally with an emotional welfare

[90] TNA MAF 369/204 Tom Ewer, "Farm animals in the law", *New Science* (18.10.1973), 179, enclosed in: Minute A. Foreman to FAWAC (08.11.1973).

[91] Tristram Beresford, "Export of livestock", *Times*, 28.03.1973, 19.

[92] TNA MAF 369/204 Ruth Harrison to Prof Hewer (13.01.1974).

[93] TNA MAF 369/204 Ruth Harrison to Prof Hewer (13.01.1974).

[94] TNA MAF 369/204 Ruth Harrison to Prof Hewer (13.01.1974).

[95] TNA MAF 369/204 Ruth Harrison to Prof Hewer (13.01.1974).

[96] TNA MAF 369/204 Prof Hewer to John [Jotcham] (17.02.1974).

lobby."[97] This did not mean that MAFF officials were unconcerned about developments. Acknowledging his ministry's failure to sponsor relevant research, MAFF's veterinary assessor for the FAWAC warned, "when we get to the Standing Committee we from the UK will have very little to offer in the way of behavioural and welfare research as compared with the Germans, the Swedes and the Dutch."[98] The UK's backwardness regarding official welfare expertise was "now a matter of fact—not of opinion."[99]

Attempting to improve the international reputation of British farming, MAFF decided to revamp FAWAC. In May 1974, Prof. Richard John Harrison from Cambridge's School of Anatomy took over chairmanship from Humphrey Hewer, who died later that month.[100] He soon proposed major reforms. Meeting MAFF officials in early 1975, Prof. Harrison criticised FAWAC for narrowly prioritising physiological data and announced a change of focus: "hitherto there had been too much emphasis on the negative aspects of welfare."[101] If FAWAC was to make "a meaningful contribution to progress,"[102] a stronger emphasis would have to be placed on animal behaviour research. FAWAC work was also being hampered by the committee's membership, which reflected industry and welfare interests rather than scientific expertise, as well as by slow information retrieval on the part of MAFF.[103]

Although Prof. Harrison's reform proposals took into account longstanding criticism by Ruth Harrison and her allies, distrust and dissatisfaction among FAWAC members remained high. Making frequent requests for field trips and information,[104] Ruth Harrison in particular continued to

[97] TNA MAF 369/204 Minute J.N. Jotcham to G.B. Taylor (18.02.1974).

[98] TNA MAF 369/204 Minute G.B. Taylor to J.N. Jotcham (20.02.1974).

[99] TNA MAF 369/204 Minute G.B. Taylor to J.N. Jotcham (20.02.1974).

[100] TNA MAF 369/204 Minute A. Foreman to FAWAC (16.05.1974); TNA MAF 369/204 Minute A. Foreman to FAWAC (31.05.1974).

[101] TNA MAF 369/206 A. Foreman, 'Note of Meeting with Professor Harrison—06.02.1975' (10.02.1975).

[102] TNA MAF 369/206 A. Foreman, 'Note of Meeting with Professor Harrison—06.02.1975' (10.02.1975).

[103] TNA MAF 369/208 A. Foreman, 'FAWAC. Meeting 25.11.1975, Supplement to Chairman's Brief (20.11.1975); TNA MAF 369/215 FAWAC, Minutes of the Fourteenth Meeting of the Committee (25.11.1975), 5–6.

[104] TNA MAF 369/208 FAWAC, Minutes of 13th Meeting (08.05.1975), 3 & 8; TNA MAF 369/215 FAWAC, Minutes of 14th Meeting of the Committee (25.11.1975), 3 & 7; TNA MAF 369/215 Minute WT Jackson to A. Foreman (12.05.1976); FAWAC General Purposes Sub-Committee, Extract from Minutes of Third Meeting (12.05.1976).

relentlessly pursue her strategy of avoiding consensus on 'weak' reforms. While this strategy was occasionally rewarded, as with the eventual revision of fowl codes for battery systems,[105] it did not win her friends. Indicating the degree of mutual distrust within FAWAC, Harrison tried to prevent the circulation of certain scientific papers amongst FAWAC members[106] and privately asked MAFF for "curriculum vitas on FAWAC members"[107] in 1976. Forwarding her request to Prof. Harrison, MAFF official Arthur Foreman replied:

> I should have thought that Members are generally adequately informed about each other's interests, and where they are not they would be able by direct enquiry to elicit any further details they would like to have.[108]

Almost ten years after the formation of FAWAC, Harrison's determination to stymie the passage of weak welfare standards had thus produced mixed results. In a far cry from the moderate image she had originally crafted to convince MAFF to nominate her for FAWAC, Harrison had prevented the rubberstamping of industry-friendly codes and successfully pushed for FAWAC's inclusion of behavioural expertise and an emerging focus on positive welfare. However, her actions also helped paralyse Britain's foremost welfare committee and prevented the passage of even marginally improved codes for many farm animals. In the absence of significant official reform, debates over welfare standards moved from the 'backstage' of Whitehall committee rooms into the 'frontstage' of the public sphere. Mirroring a wider fraying of post-war corporatism, the 1970s were characterised by increasingly polarised public clashes over farm animal welfare between producers and campaigners—and between campaigners themselves. For Harrison, this transition would prove difficult. While MAFF officials and many members of FAWAC increasingly viewed her as a radical, younger activists considered her advocacy for gradualist reforms of intensive farming as too moderate.

[105] TNA MAF 369/272 Background to Question 6, enclosed in: Handwritten list of questions and answers to enquiry by House of Commons Agriculture Committee, 3–4.

[106] TNA MAF 369/215 Minute [illegible] to A. Foreman (08.06.1976).

[107] TNA MAF 369/215 A. Foreman to Ruth Harrison (18.06.1976), 2, enclosed in: A. Foreman to Professor Harrison (18.06.1976), 2.

[108] TNA MAF 369/215 A. Foreman to Ruth Harrison (18.06.1976), 2, enclosed in: A. Foreman to Professor Harrison (18.06.1976), 2.

BY

CHAPTER 9

Ruth the Ruthless: Activism, Welfare, and Generational Change

The polarisation of 1970s' welfare debates triggered an increasing professionalisation of British animal campaigning as well as generational clashes about the style and goals of protest. The described changes were not unique to animal campaigning. Since the 1950s, many non-governmental organisations (NGOs) had begun to professionalise their organisational structures and lobbying tactics, and environmentalist organisations like Greenpeace had pioneered new forms of direct action protest.[1] The changing campaigning environment created pressure on the traditionalist RSPCA leadership to rethink its own tactics. Since the mid-1960s, the Society's Council had tried to shape emerging farm animal welfare debates with a mix of traditional backstage lobbying, strategic sponsorship of scientific research and expertise, and investment in expensive media campaigns. This strategy resulted in a number of political victories. However, the Council's parallel tendency to quell internal conflicts by expelling vocal critics and ongoing ambivalence over so-called field sports (Chap. 5) alienated many RSPCA members. Between 1970 and 1977, escalating tensions over hunting, intensive farming, and Council decisions resulted in a rise to power of the so-called RSPCA Reform Group. Members of this group oversaw sweeping organisational reforms and popularised

[1] Zelko, *Make It a Green Peace*; Matthew Hilton, James McKay, Nicholas Crowson, and Jean-François Mouhout, *The Politics of Expertise: How NGOs Shaped Modern Britain* (Oxford: Oxford University Press, 2013), 81–100; Nixon, "Trouble at the National Trust".

C. Kirchhelle, *Bearing Witness*, Palgrave Studies in the History of Social Movements,
https://doi.org/10.1007/978-3-030-62792-8_9

contemporary animal rights thinking. At its highpoint, reform leadership also threatened the Society's traditionally close ties to Britain's political establishment. As a Council member, Ruth Harrison played an important role both in kick-starting the RSPCA's engagement with farm animal welfare and in triggering the internal crisis that led to the rise of the Reform Group.

Becoming a Professional

Harrison's engagement with professional animal activists outside of government committees had begun unsystematically. In the years following the publication of *Animal Machines*, she spoke at numerous conferences on 'factory farming'[2] and also engaged in occasional publicity stunts. In 1968, she accepted a challenge from intensive farmer Alistair Nugent, nephew of Conservative politician and former NFU vice-chairman Lord Nugent of Guildford. Together with Nugent, she devised a questionnaire on 'factory farming' that was used to see if Nugent could convert 40 opponents of intensive agriculture during a visit to Nugent's farm. In the end, Nugent failed to convert a single opponent. According to the *Times*, "one woman cried when she saw the caged hens."[3]

Harrison also campaigned against the slaughter of non-stunned animals. Together with Muriel Dowding, founder of the *Beauty Without Cruelty* charity for eliminating cosmetics testing on animals, she asked Labour's Baroness Edith Summerskill to support a proposed ban on the 'ritual' slaughter of non-stunned animals in 1966. However, Baroness Summerskill would support the motion only if Harrison and Dowding managed to secure a promise "from the Imam that the Moslems will agree to stunning."[4] Such a promise was forthcoming from neither Muslim nor Jewish communities. According to Dowding, politicians on both sides of the aisle were afraid to support bans for fear of losing votes.[5] One of the few politicians to support a ban in the House of Lords was Dowding's

[2] "Danger in antibiotic spread", *Times*, 24.08.1968, 14.

[3] "Hen 'factory' makes no converts", *Times*, 12.07.1968, 4.

[4] FACT Files, DB, Unmarked Green Ryman Folder, Lady Dowding to Ruth Harrison (21.09.1966); see also FACT Files, DB, Unmarked Red Ryman Folder, Ruth Harrison to David (25.02.1968) [second date: 11.06.1971].

[5] FACT Files, DB, Unmarked Green Ryman Folder, Dowding to Harrison (21.09.1966).

husband, Commander of the Battle of Britain, Lord Dowding, who was now a fervent if isolated animal protection advocate.[6]

Harrison's failure to mobilise support for a reform of non-stun regulations showed that relying on her bestseller fame would not be enough to shape the trajectory of British animal politics. To become more effective as a campaigner, she would have to find additional financial, logistical, and expert support for her work. One way of doing so was to create her own charity. In the same month that she began to lobby for FAWAC membership, Ruth Harrison started to approach renowned scientists and potential sponsors for the formation of a new research trust devoted to animal welfare.[7] Amongst the confirmed trustees was Cambridge veterinary researcher and frequent Harrison critic David Sainsbury.[8] Other potential trustees were Prince Phillip and E.F. Schumacher, renowned economic adviser of the National Coal Board, leading member of the *Soil Association*, and later author of *Small Is Beautiful* (1973).[9] However, despite speaking against factory farming at a conference organised by the Ruth Harrison Advisory Group at London's Friends' House in 1966,[10] Schumacher did not become a trustee. The Trust itself was registered in early 1967 as the Ruth Harrison Welfare Trust but was hastily renamed as the Ruth Harrison Research Trust to distinguish itself from general welfare trusts.[11]

Generating funds for the new Ruth Harrison Research Trust proved more difficult than expected. In January 1968, trustees met for the first time. Ruth Harrison was elected Chairman of the Trust, and it was resolved to raise "a fighting fund" by making approaches "to possible wealthy sources."[12] Despite attracting donors like Lord Conesford and the Whitley

[6] FACT Files, DB, Unmarked Green Ryman Folder, Dowding to Harrison (21.09.1966); on the Dowdings see Roscher, *Königreich*, 247.

[7] FACT Files, DB, Appelbe, Ambrose Appelbe to Dr Harrison (19.08.1966).

[8] FACT Files, DB, Appelbe, Ruth Harrison to Ambrose Appelbe (23.08.1966).

[9] FACT Files, DB, Appelbe, Ruth Harrison to Ambrose Appelbe (23.08.1966); Ruth Harrison to Ambrose Appelbe (02.02.1967); Veldman, *Fantasy, the Bomb and the Greening of Britain. Romantic Protest, 1945–1980*, 292–99.

[10] Stanley Baker, "Factory farms 'no answer'", *Guardian*, 17.10.1966, 4.

[11] FACT Files, DB, Appelbe, Appelbe to Charity Commission (17.02.1967); Harrison to Appelbe (25.07.1967); the Research Trust only seems to have been renamed Farm Animal Care Trust in 1974; "Public Notices—Charity Commission", *Times*, 13.12.1974, 26.

[12] FACT Files, MD, Minute Book, Ruth Harrison Research Trust, Minutes of 1st Meeting of Trustees, 02.01.1968, 2.

Animal Protection Trust,[13] there was little steady income. In 1967, one supporter sagely noted, "you are going to have very little money for a very long time."[14] With *Animal Machines* fading from the public's mind, it was also becoming clear that naming the trust the Ruth Harrison Research Trust had been a mistake.[15] Discussing the issue, a correspondent noted: "If I hadn't read your book, I shouldn't want to give a penny to a trust with the name you quoted."[16] In 1970, David Sainsbury expressed similar concerns about the "unfortunate confusion that exists with the name of the 'Ruth Harrison Research Trust'":

> the implication to the 'reader' is that this body is a distributor of funds to bodies doing work we approve of … For some reason I find they do not associate this name with a body eagerly seeking funds, as well as promoting and, we hope, financing research.[17]

Some also believed that by focusing on food quality as well as on farm animal welfare, the Trust's interests were too broad.[18] As a consequence, it was decided to abandon an explicit focus on food quality and look for a new name that more closely associated the Trust with farm animal welfare. Although Harrison remained interested in environmental and nutritional issues,[19] the Ruth Harrison Research Trust was renamed as the Farm Animal Care Trust (FACT) in October 1974.[20] In the preceding interim

[13]FACT Files, DB, Unmarked Green Ryman Folder, Lord Conesford to Ruth Harrison (08.12.1969); Mr Burns to Ruth Harrison (13.03.1969).

[14]FACT Files, DB, Unmarked Red Ryman Folder, Kenneth to Ruth Harrison (10.03.1967), 4.

[15]FACT Files, DB, Unmarked Red Ryman Folder, Kenneth to Ruth Harrison (10.03.1967).

[16]FACT Files, DB, Unmarked Red Ryman Folder, Kenneth to Ruth Harrison (10.03.1967), 3.

[17]FACT Files, DB, Dr Sainsbury, David Sainsbury to Ruth Harrison (08.07.1970).

[18]FACT Files, DB, Unmarked Green Ryman Folder, Ruth Harrison to Mr Lustgarten (24.11.1967); FACT Files, DB, Unmarked Red Ryman Folder, Kenneth to Ruth Harrison (10.03.1967), 4.

[19]Kendall, "Ruth and the Ruthless," 2; Dex Harrison was also interested in the interface of architecture and animal welfare; FACT Files, DB, Unmarked Blue Ryman Folder, Article, "Farm Fires—A National Scandal," *Architects' Journal* 41/156 (12.10.1972).

[20]FACT Files, MD, Minute Book, Ruth Harrison, FACT, Minutes of Meeting of the Trustees held on 09.05.1978 (14.05.1981), 2; FACT Files, Appelbe, Ambrose Appelbe and Partners to Ruth Harrison (29.11.1971); FACT Files, Marian Stamp Dawkins [in the following MD], Minute Book, Farm Animal Care Trust. Meeting of Trustees (11.08.2000).

phase, pamphlets already pointed to the Trust's scientific focus on farm animal welfare as its distinguishing feature:

> Because farming systems and techniques change so rapidly, a specialist body, the Ruth Harrison Trust, has been formed to initiate further research and maintain an adequate educational programme for the public at large. IT IS THE ONLY RESEARCH ORGANISATION WHOSE ENTIRE ACTIVITY IS DEVOTED TO FARM ANIMALS [sic].[21]

Despite its meagre resources, Harrison's Trust soon funded nutritional research at Queen Elizabeth College and research on improving stunning techniques and poultry transports.[22] Following the election of former Brambell member William Homan Thorpe as Trustee in February 1968,[23] Harrison also commissioned a Gallup Poll of farmers' attitudes towards 'factory farming' and husbandry systems.[24] In total, Gallup conducted ca. 1900 interviews with farmers from all over Britain. The results seemed to confirm Harrison's assertion that intensive systems were controversial amongst farmers themselves:

FACT Files, DB, Farming Survey, Enclosed in: HB Fawcett to FAWAC members, 'FAWAC. Gallup Poll Result" (Office Note 14.02.1969), 4.

Total 1900 *Farmers agreeing with*	*Per cent*
Complete diet	94
Sufficient light	95
Access to daylight	72
Free movement	85
Room for wings	78
Comfort of floors	89
Access to outdoor for cattle-sheep	76
Access to outdoor for other stock	56

[21] FACT Files, DB, Unmarked Green Ryman Folder, Draft Model Pamphlet—The Ruth Harrison Trust, 1.

[22] FACT Files, MD, Minute Book, Ruth Harrison Research Trust, Minutes of 1st Meeting of Trustees, 02.01.1968, 2; FACT Files, D.B., Unmarked Red Ryman Folder Ruth Harrison to David Sainsbury (27.10.1968).

[23] FACT Files, MD, Minute Book, Ruth Harrison Research Trust, Resolution of the Trustees (20.02.1968).

[24] FACT Files, DB, Unmarked Blue Ryman Folder, Farming Survey, Enclosed in: H.B. Fawcett to FAWAC members, 'FAWAC. Gallup Poll Result'. Office Note (14.02.1969).

An Uneasy Relationship: Ruth Harrison and the RSPCA

Being able to commission opinion polls and supportive research was becoming increasingly important for Harrison's work within FAWAC (Chap. 8). However, her strategy of blocking FAWAC compromises on weak welfare codes also necessitated generating sufficient external public and political pressure to break regulatory deadlocks in favour of more ambitious welfare measures. Forming an alliance with an established campaigning organisation was one way of doing so. Its prestige, financial power, and corporatist ties made the RSPCA an obvious choice.

Harrison had been a member of the RSPCA since 1964, and her appointment to the RSPCA's Council in April 1969 held mutual advantages.[25] Harrison stood to profit from the Society's close relations to the Parliamentary Group for Animal Welfare and ability to commission large-scale publicity campaigns. In turn, the RSPCA Council profited from Harrison's public prestige as the author of *Animal Machines* and her insider knowledge as the most high-profile welfarist FAWAC member.[26] Strategic goals aligned closely. Since the publication of *Animal Machines*, the RSPCA had supported campaigns against battery cages for poultry, crates and deficiency diets in veal production, cattle docking, and weak welfare provisions in the 1968 Agriculture (Miscellaneous Provisions) Bill.[27] It had also intensified campaigns to outlaw the live export of British animals to foreign slaughterhouses[28] as well as against Jewish and Muslim "ritual slaughter"[29] without pre-stunning.

[25] RSPCA Archives, IF/56/6, Ruth Harrison General File, RM/A853, document: Harrison—Mrs Ruth, OBE 1988.

[26] The Council had not been consulted by MAFF about the nomination of Irene Walsh and was eager to expand its influence within FAWAC, RSPCA Archives, CM/57 RSPCA Council Minutes 1966–1968, Meeting of the Council, 27.07.1967, 4.

[27] RSPCA Archives, CM/55 RSPCA Council Minutes 1962–1966, Meeting of the Council, 21.01.1965, 2–3; 10.06.1965, 2; 17.02.1966, 2–3; CM/57 RSPCA Council Minutes 1966–1968, Meeting of the Council, 23.11.1967, 3; 25.07.1968, 2; Meeting of Council, 25.07.1968, 2.

[28] RSPCA Archives, CM/55 RSPCA Council Minutes 1962–1966, Meeting of the Council, 15.07.1965, 2; 21.10.1965, 2; CM/57 RSPCA Council Minutes 1966–1968, Meeting of the Council, 27.07.1967, 4; CM/58 RSPCA Council Minutes 1968–1970, Meeting of the Council, 24.10.1968, 4, 6–7; Meeting of the Council, 27.02.1969, 2–3.

[29] RSPCA Archives, CM/57 RSPCA Council Minutes 1966–1968, Meeting of the Council, 23.05.1967, 3–4; Meeting of the Council, 25.04.1967, 4–5; Meeting of Council,

Harrison's appointment triggered a flurry of RSPCA farm animal welfare activities. At her second Council meeting in May 1969, she re-invigorated the RSPCA's defunct sub-committee on intensive methods of animal husbandry, which had been founded in 1964.[30] At her third meeting, she proposed a resolution for the Society's upcoming general assembly according to which the RSPCA deplored battery systems in general but—pending the ban of such systems—specifically opposed the dewinging of birds.[31] And at her fourth meeting, she highlighted inaccuracies in RSPCA material on animal welfare, gained a seat on the general purposes sub-committee and the new ad hoc committee on intensive farming, and successfully called on the RSPCA to establish legal precedent by securing cruelty prosecutions against intensive farms.[32]

In tandem with Harrison, the RSPCA's leadership also devised a publicity campaign to prevent the parliamentary enactment of weak initial FAWAC codes "by means of press advertising and the widespread distribution of literature."[33] Harrison accompanied the RSPCA campaign with media broadcasts and newspaper articles. In Parliament, Conservative MP and RSPCA Council member Frank Burden tabled an "amendment regretting that the codes failed to implement the recommendations of the Brambell Committee and requesting the Government to introduce amended codes in the next session."[34] By the end of October 1969, the collaboration between Britain's most senior animal protection

25.07.1968, 6; for parallel controversies between liberal and orthodox parts of Britain's Jewish community see, RSPCA Archives, CM/55 1962–1966, RSPCA Council Minutes, Meeting of the Council, 11.06.1964, 3; CM/58 RSPCA Council minutes 1968–1970, Meeting of the Council, 27.02.1969, 1.

[30] RSPCA Archives, CM/58 RSPCA Council Minutes 1968–1970, Meeting of the Council, 22.05.1969, 6; CM/55 RSPCA Council Minutes 1966–1968, Meeting of the Council, 21.05.1964, 3.

[31] RSPCA Archives, CM/58 RSPCA Council Minutes 1968–1970, Special Meeting of the Council, 05.06.1969, 1.

[32] RSPCA Archives, CM/58 RSPCA Council Minutes 1968–1970, Meeting of the Council, 26.06.1969, 2–3, 5; Harrison also tried to stop the Royal Agricultural Show from hosting a rodeo; RSPCA Archives, CM/58 RSPCA Council Minutes 1968–1970, Meeting of the Council, 27.11.1969, 2.

[33] RSPCA Archives, CM/58 RSPCA Council Minutes 1968–1970, Meeting of the Council, 23.10.1969, 1.

[34] RSPCA Archives, CM/58 RSPCA Council Minutes 1968–1970, Meeting of the Council, 23.10.1969, 2.

organisation and most prominent farm animal welfare campaigner had resulted in a rare MAFF promise to review codes (see Chap. 8).

The honeymoon between Harrison and the RSPCA Council was brief. Probably buoyed by her recent success, Harrison triggered a significant rift among Council members in early 1970 by leaking a confidential letter on 'field sports.' The letter marked the most recent escalation of attempts by the British Field Sports Society (BFSS) to influence RSPCA hunting policies. After reaching a high around 1961, the mid-1960s had seen internal RSPCA tensions over hunting simmer down following the expulsion of radical anti-hunt activists like Patrick Moore, Howard Johnson, and Gwendolen Barter (Chap. 5). In public, the RSPCA would voice concerns about hunting for sport but would take little concrete action.[35] In November 1968, senior RSPCA executives even met with BFSS head and Conservative MP Marcus Kimball over lunch. During the meeting, the BFSS proposed agreeing on public talks on "improving standards in shooting and fishing."[36] Although the RSPCA rejected this proposal, it later agreed to a public "statement about ongoing talks" and the "hope 'for a better understanding of each other's position' in future."[37]

While relations between the RSPCA leadership and BFSS seemed to be improving, those with hunt critics were deteriorating. Since the early 1960s, the League Against Cruel Sports (LACS) had intensified campaigning for RSPCA condemnations of 'field sports' like fox hunting with hounds or hare coursing at general meetings and with mass letters to RSPCA members. This behaviour threatened the authority of the Council. In May 1969, RSPCA Chairman John Hobhouse condemned "interference" by LACS circulars in the RSPCA's postal election and "expressed concern that, seemingly, matters raised in confidence at Council Meetings had been imparted to the League."[38] However, LACS advocacy proved popular among RSPCA members and parliamentary supporters. In the summer of 1969, 100 MPs filed a motion recognising the "overwhelming public support" for bans and urged the government "as a contribution to

[35] Ryder, *Animal Revolution*, 171–173; Roscher, *Königreich*, 290–291.

[36] RSPCA Archives, CM/58 RSPCA Council Minutes 1968–1970, Meeting of the Council, 27.02.1969, 5.

[37] RSPCA Archives, CM/58 RSPCA Council Minutes 1968–1970, Meeting of the Council, 27.02.1969, 6; Meeting of the Council, 28.11.1968, 4.

[38] RSPCA Archives, CM/58 RSPCA Council Minutes 1968–1970, Meeting of the Council, 22.05.1969, 6.

the European Conservation Year" to "introduce appropriate legislation."[39] Labour MP Arnold Shaw also filed a private members bill urging bans of hare coursing and hunting wild deer with hounds. RSPCA Council members like Ruth Harrison campaigned for the Society to support Shaw's bill in line with a recent general meeting resolution to "do all in [the Society's] power to end deer hunting and hare coursing."[40]

However, to critics' dismay, the RSPCA failed to publicly support the Bill. Behind the scenes, this inaction had been caused by a BFSS letter threatening to challenge the RSPCA's charity status if it "actively campaigned against coursing and deer hunting."[41] Losing charity status could have cost the RSPCA up to £300,000 annually.[42] In November 1969, "traditionalist"[43] MP Frederick Burden, who had just been elected as RSPCA vice-chairman,[44] explained the threat to Ruth Harrison but stressed that the BFSS letter must remain confidential while the RSPCA assured itself of its legal status.[45] Unwilling to wait, Harrison mentioned the letter at an RSPCA Council meeting on November 27. However, a majority voted to delay campaigning and approaching UK Charity Commissioners until legal clarity had been obtained.[46]

Following this delay, things escalated rapidly: in early December, Harrison mentioned the BFSS letter to the *Guardian* and claimed to be "deeply disappointed that by its inactivity the [RSPCA] appears to have taken this lying down, ... I think it best to bring the whole business into the open."[47] At a Council meeting on December 31, she denied that there

[39] RSPCA Archives, CM/58 RSPCA Council Minutes 1968–1970, Meeting of the Council, 24.07.1969, 1; anti-hunt protest was also supported by newspapers like the *Times* "Outdated: Objectionable", *Times*, 05.11.1969, 9.

[40] "Blackmail over blood sports, says RSPCA", *Guardian*, 11.12.1969, 22.

[41] "Blackmail over blood sports, says RSPCA", *Guardian*, 11.12.1969, 22.

[42] "Charity fears £300,000 fall in income", *Guardian*, 17.05.1973, 7; according to Richard Ryder, the fears were a farce and used to stop internal opposition, Ryder, *Animal Revolution*, 173.

[43] Ryder, *Animal Revolution*, 175.

[44] RSPCA Archives, CM/58 RSPCA Council Minutes 1968–1970, Meeting of the Council, 26.06.1969, 2; Meeting of the Council, 24.07.1969, 3; Meeting of the Council, 23.10.1969, 4.

[45] "Charity fears £300,000 fall in income", *Guardian*, 17.05.1973, 7.

[46] RSPCA Archives, CM/58 RSPCA Council Minutes 1968–1970, Meeting of the Council, 27.11.1969, 1, 7–8; "Jury in RSPCA case says member's letter was not a libel", *Guardian*, 19.05.1973, 6.

[47] "Blackmail over blood sports, says RSPCA", *Guardian*, 11.12.1969, 22; see also: "Charity fears £300,000 fall in income", *Guardian*, 17.05.1973, 7; "Jury find that ballerina's letter was true", *Times*, 19.05.1973, 2.

had been a "breach of confidence on Council matters, but only of a telephone conversation she had had with Mr Burden, to whom she offered a full apology."[48] Belying later descriptions of her as giving "the impression of a certain fragility,"[49] Harrison felt that:

> the [RSPCA] chairman should have found out from [Legal] Counsel at the meeting on 17th November, what positive action could be taken to support Mr Arnold Shaw's Bill and that as she considered that the RSPCA did not intend to take any useful action, she could best help the cause of animal welfare and those in Parliament by disclosing the threat by the BFSS.[50]

RSPCA Chairman John Hobhouse countered that details of Shaw's Bill had only been made public in early December, which would have precluded organising a full publicity campaign. A supportive letter to the *Times* by himself had not been published.[51] In response, Harrison pushed for the expulsion from the RSPCA of the four BFSS members, among them Marcus Kimball, who had threatened to challenge the Society's charity status.[52] This suggestion met with legal objections, and the Council only agreed to explore whether subscriptions could be cancelled.

In view of this further delay, Harrison supplied a confidential photocopy of the original BFSS letter, which she had obtained from Hobhouse on December 31, to LACS chairman Raymond Rowley. Rowley broke his promise to only use the photocopy to obtain a second legal opinion and used it to disrupt the BFSS annual meeting. When asked by the RSPCA whether she had leaked the letter, Harrison refused to confirm or deny allegations and failed to appear at relevant Council meetings.[53] This behaviour infuriated other Council members. Elected to the RSPCA Council

[48] RSPCA Archives, CM/58 RSPCA Council Minutes 1968–1970, Meeting of the Council, 31.12.1969, 3.

[49] Kendall, "Ruth and the Ruthless," 2.

[50] RSPCA Archives, CM/58 RSPCA Council Minutes 1968–1970, Meeting of the Council, 31.12.1969, 3.

[51] RSPCA Archives, CM/58 RSPCA Council Minutes 1968–1970, Meeting of the Council, 31.12.1969, 3.

[52] RSPCA Archives, CM/58 RSPCA Council Minutes 1968–1970, Meeting of the Council, 31.12.1969, 9.

[53] RSPCA Archives, CM/59 RSPCA Council Minutes 1970–1971, Meeting of the Council, 30.07.1970, 5; see also, British Library, Richard Ryder Papers, RSPCA Ryder Dep. 9856, B2/2, RSPCA Reform, 1971–1972, 2. 1974–75, RSPCA Reform Group News Letter, February 1974—Ruth Harrison, April 1974.

alongside Harrison in March 1969,[54] famed ex-Prima Ballerina Nadia Nerina described Harrison's evasions as "absolutely disgraceful."[55] In a circular letter to Council members from November 1970,[56] Nerina suggested that Harrison was "not fit to be a member of the RSPCA."[57] Reacting to this letter and growing criticism of her in the Council, Harrison decided to sue Nerina for libel in 1972. The libel suit was a grave miscalculation. In May 1973, Harrison lost her suit and was ordered to pay ca. £30,000 in court costs.[58] In his verdict, the presiding judge ruled that there was "no question of malice in Miss Nerina's actions" and that it was a tragedy that "two women of worth, devoted to animal welfare," should find themselves "at arm's length over their concern for animals."[59]

Over the following months, prominent supporters like Yehudi Menuhin, Julian Huxley, the Archdeacon of Westminster, and Dame Margery Perham established a fund to pay for Harrison's legal costs and succeeded in raising £6000. Meanwhile, Harrison announced that she was taking up a job to pay for the court costs.[60] Referring to her case, RSPCA critics like the *Guardian*'s Martin Walker accused the Society of misusing charitable funds to legally silence critical members.[61] Harrison herself poured further fuel into the fire in 1974 by revealing that a new RSPCA-promoted humane electric stunner did not conform to British standards.[62] Nonetheless, all attempts to cover the court costs proved futile. On June 12, 1975, the *London Gazette* reported that Ruth Harrison, "(married woman), of 34, Holland Park Road, London, ..., occupation unknown,"[63] had filed a petition for bankruptcy, which had now been proven. Harrison's reaction to the bankruptcy is telling. Rather than yield her seat on the RSPCA Council, she publicly defended her actions:

[54] "Elected to Council of RSPCA", *The Glasgow Herald*, 09.04.1969, 9.

[55] "Charity fears £300,000 fall in income", *Guardian*, 17.05.1973, 7.

[56] "Jury in RSPCA case says member's letter was not a libel", *Guardian*, 19.05.1973, 6.

[57] "Jury find that ballerina's letter was true", *Times*, 19.05.1973, 2.

[58] Kendall, "Ruth and the Ruthless," 21.

[59] "Jury in RSPCA case says member's letter was not a libel", *Guardian*, 19.05.1973, 6.

[60] Kendall, "Ruth and the Ruthless," 21; Edward Carpenter, "Ruth Harrison", *Times*, 27.07.1973, 17; "Debt of honour", *Observer*, 02.09.1973, 40.

[61] Martin Walker, "Open file—Doggy fashion", *Guardian*, 10.01.1974, 13.

[62] "Humane Killer—'inhumane'", *Observer*, 15.09.1974, 1–2.

[63] "The Bankruptcy Acts 1914 and 1926—Receiving Orders", *The London Gazette* (12.06.1975), 7612.

> My first loyalty must be to the cause for which I was elected, my second to the members who elected me, my third to the Council itself. At no time would I promise confidentiality at the risk of suffering to animals.[64]

A New Style of Activism: The Rise of the RSPCA Reform Group

Harrison's actions ended the RSPCA Council's detente on hunting.[65] Mirroring a wider turn away from the "softly-softly"[66] reformism of the 1960s, the 1970s saw younger activists shake up British civic activism in fields ranging from gay rights to environmentalism. In the case of the RSPCA, the BFSS episode made growing tensions between older 'traditionalists' in the Council and younger grassroots activists boil over. Founded in 1970, the RSPCA Reform Group criticised an allegedly elitist leadership for ignoring majority demands for decisive action against 'field sports.' At stake was not just the issue of hunting but a wider revaluation of Council accountability and internal democracy.

Ahead of the Society's 1970 annual meeting, RSPCA Chairman John Hobhouse and Vice-Chairman Frank Burden attempted to diffuse tensions with a referendum on whether 'field sports' should be discussed. Of polled members 4028 voted against and 3836 in favour, and all resolutions regarding 'field sports' were disallowed.[67] The referendum failed to mollify critics, and the RSPCA's June 1970 annual meeting had to be closed early due to disruptions. Parts of the increasingly divided RSPCA Council subsequently tried to quell protest by expelling disruptive members and threatening legal action.[68] This approach backfired.

Emboldened by the referendum and supported by the National Society for the Abolition of Cruel Sports, the Reform Group tried to gain a foothold in the Council by vetting candidates for potential hunting affiliations and supporting campaigns of allied activists.[69] In a letter to RSPCA

[64] Quoted according to Kendall, "Ruth and the Ruthless," 21.

[65] Ryder, *Animal Revolution*, 171–173; Roscher, *Königreich*, 295–298.

[66] Lent, *British Social Movements*, 97; Roscher, *Königreich*, 260–273.

[67] RSPCA Archives, CM/59 RSPCA Council Minutes 1970–1971, Special Meeting of the Council, 04.06.1970, 1 & 4.

[68] RSPCA Archives, CM/59 RSPCA Council Minutes 1970–1971, Meeting of the Council, 30.07.1970, 7; Meeting of the Council, 28.01.1971, 7–8; "New attack on RSPCA planned", *Guardian*, 22.06.1970, 6.

[69] British Library, Richard Ryder Papers, Ryder Dep 9846 B2; 1, RSPCA Reform, 1971–1975, National Society for the Abolition of Cruel Sports to RSPCA Members, 14.04.1972.

employees, Reform Group co-founder John Bryant claimed "that hunting people … had systematically plotted to gain control of the RSPCA."[70] Employees should think through "facts" and "show, which Council members are biased": "you must not hesitate to rid the Society of any Council member who tolerates a particular form of cruelty."[71] RSPCA leadership reacted with a carrot and stick policy. In Spring 1971, the Council passed a new policy statement on hunting according to which the "RSPCA deplores the unnecessary killing of any wild creature or the infliction of avoidable suffering and distress."[72] It also considered assessing whether wild animals could be brought within the provisions of the 1911 Protection of Animals Act and renewed enquiries into humane alternatives to fox hunting with hounds. Ahead of the 1971 General Meeting, Chairman Hobhouse also met with Reform Group members and promised an anti-blood sports motion in return for a no-disruption guarantee.[73] Reform Group members adapted their tactics. In preparation for the general meeting, notes informed Reform Group members to "act orderly," "disburse" in the crowd, and use points of order to challenge "bloodsports men": "Do not show this note to anyone else. If they haven't a copy, they're not one of us."[74]

With activists continuing to push for RSPCA condemnations of specific hunting practices like fox hunting with hounds,[75] tensions escalated further. Ahead of the 1972 postal votes for Council membership, it seemed likely that the Reform Group would launch a letter campaign to influence voting.[76] After considerable discussion, the Council agreed to counter Reform Group views with an explanatory letter that could be sent

[70] British Library, Richard Ryder Papers, Ryder Dep. 9846 B2;1 RSPCA Reform, 1971–1975, John Bryant to RSPCA Officials, 08.02.1971, 1.

[71] British Library, Richard Ryder Papers, Ryder Dep. 9846 B2;1 RSPCA Reform, 1971–1975, John Bryant to RSPCA Officials, 08.02.1971, 1.

[72] RSPCA Archives, CM/59 RSPCA Council Minutes 1970–1971, Meeting of the Council, 03.03.1971, 5.

[73] RSPCA Archives, CM/59 RSPCA Council Minutes 1970–1971, Meeting of the Council, 03.03.1971, 5–6, 10–11.

[74] British Library, Richard Ryder Papers, Ryder Dep. 9846 B2;1 RSPCA Reform, 1971–1975, RSPCA Reform Group: Note For Supporters. Re RSPCA AGM; prepared on evening of 24.06.1971.

[75] RSPCA Archives, CM/59 RSPCA Council Minutes 1970–1971, Meeting of the Council, 05.08.1971, 12–13.

[76] RSPCA Archives, CM/60 RSPCA Council Minutes 1971–1972, Meeting of the Council, 05.04.1972, 7–8.

alongside voting papers. Because Reform Group membership was a secret, it was also decided to enhance CV requirements for Council candidates to hinder infiltration.[77] A motion to ask the Reform Group to disclose its membership was also debated.[78] However, despite these measures, the 1972 Council elections returned not only existing anti-hunt members like Harrison but also Reform Group members like John Bryant, Bryan Seager, Andrew Linzey, and Richard Ryder.[79]

With much of recent RSPCA historiography either written by former Reform Group members or drawing on their accounts, the 1972 elections have been described as the beginning of a marked break in the Society's history.[80] This is an exaggeration. The election of Bryant, Seager, Linzey, and Ryder was certainly a triumph for the Reform Group. However, Council voting patterns and the continued election of many "traditionalists" clearly show that a majority of the RSPCA's leadership, Council, and membership remained "traditionalist." Rather than marking a revolution, the 1972 elections accelerated organisational transformations that were already taking place. These transformations were occurring not just in response to the much-publicised tensions over hunting but against a background of rapidly increasing demands on the RSPCA and a wider post-war professionalisation of NGOs. According to Matthew Hilton and others, this professionalisation was characterised by "the focused, professional pursuit of fundraising, marketing and advertising" and driven by the increasing importance of the "marshalling of expertise."[81]

With backstage decision-making on British animal welfare politics breaking down (Chap. 8), it was no longer sufficient for the RSPCA to act as a traditional charity, lobby politicians, police individual acts of cruelty, and run shelters. Instead, it had to find new ways of maintaining its influence over the rapidly expanding and increasingly crowded political

[77] RSPCA Archives, CM/60 RSPCA Council Minutes 1971–1972, Meeting of the Council, 05.04.1972, 7–8.

[78] RSPCA Archives, CM 60 RSPCA Council Minutes 1971–1972, Meeting of the Council, 07.06.1972, 7.

[79] Richard Ryder, "RSPCA Reform Group," in Marc Bekoff and Carron A. Meaney (eds.), *Encyclopedia of Animal Rights and Animal Welfare* (Abingdon/New York: Routledge, [1998] 2013), 307–308; British Library, Richard Ryder Papers, Ryder Dep. 9846 B2;1 RSPCA Reform, 1971–1975, National Society for the Abolition of Cruel Sports, 14.04.1972 to RSPCA members.

[80] Ryder, "RSPCA Reform Group," 492–493; Ryder, *Animal Revolution*, 174–177.

[81] Hilton et al., *Politics of Expertise*, 80–81.

marketplace for animal welfare, environmentalism, and conservation in the midst of an emerging fiscal crisis and Britain's accession to European policy frameworks.[82]

By 1972, this transition was well underway. Between 1970 and 1975, a 'traditionalist' Council under RSPCA Chairman Hobhouse oversaw ambitious reforms of the Society's campaigning and involvement with welfare scientists. In 1971 and 1972, the Society replaced existing ad hoc committees with expert advisory committees on animal experimentation and farming. Council members were still present, but the committees were mostly staffed with scientists and representatives of relevant professions.[83] Discussed in more detail in Chap. 10, the new Farm Livestock Advisory Committee (FLAC) began to publish detailed scientific reviews of British welfare codes.[84] For the first time in its history, the Society also began to actively sponsor scientific research and co-organised a major conference on stress in 1973.[85] Responding to growing demands for empirical data, the Society's veterinary staff conducted national and international surveys of animal welfare. Between 1972 and 1973, the RSPCA also launched a successful lobbying campaign to Stop the Export of Live Farm Animals (SELFA).[86] To support its expanding activities, the RSPCA hired new Education Officers, raised expenditure, and increased staff workloads. By 1973, headquarters staff received ca. 1000 letters, sent out ca. 700 letters—excluding mass mail outs—and answered around 300/400 telephone calls per day.[87]

[82] For an overview of other new farm animal campaigning organisations like Compassion in World Farming (est. 1967), see Roscher, *Königreich*, 260–266; 290–293.

[83] RSPCA Archives, CM/58 RSPCA Council Minutes 1968–1970, Private and Confidential—Council and Standing Committee, 23.12.1969, 1; CM/59, RSPCA Council Minutes 1970–1971, Meeting of the Council, 05.08.1971, 2; Richard Ryder, "Putting Animals into Politics," in Robert Garner (ed.), *Animal Rights. The Changing Debate* (Basingstoke and London: Macmillan, 1996), 173; Ryder, *Animal Revolution*, 175.

[84] RSPCA Archives, FLAC Minutes, Meeting 20.07.1971, 2–3.

[85] RSPCA Archives, CM/61 RSPCA Council Minutes 1972–1975; Meeting of the Council, 26.10.1972, 10–11; Kirk, "Invention of the Stressed Animal," 256; Roscher, *Königreich*, 294–97.

[86] RSPCA Archives, CM/61 RSPCA Council Minutes 1972–1975, Meeting of the Council, 26.10.1972, 4–5; Meeting of the Council, 07.02.1973, 1–2; Meeting of the Council, 03.01.1974, 2 & 8; RSPCA Archives, FLAC Minutes, Meeting 27.02.1973, 2; RSPCA Archives, CM/59 RSPCA Council Minutes 1970–1971, Meeting of the Council, 01.04.1971, 1.

[87] RSPCA Archives, CM/61 RSPCA Council Minutes 1972–1975, Statement by Executive Director at Meeting of the Council, 01.11.1973 (item 10(8(c))); "Bankruptcy warning by RSPCA chief," *Times*, 29.06.1974, 2.

Despite intensifying RSPCA welfare research and lobbying, Reform Group members continued to press for more explicit condemnations of 'field sports' and for the democratisation of leadership structures. In November 1972, Reform Group member Bryan Seager supported public protest against Princess Anne's participation in a hunting excursion at the Zetland Hunt.[88] This protest had explosive potential within a Society that prided itself on its Royal Patronage. While the RSPCA Council quickly distanced itself from Seager's protest,[89] the Reform Group escalated the situation by calling for the removal of Royal patronage should the Royal Family not distance itself from hunting activities.[90] In a sign of how polarised the situation had become, both the Reform Group's initiative and subsequent 'traditionalist' attempts to expel Seager failed.[91]

Concerned about growing Reform Group agitation, the RSPCA's leadership resorted to increasingly controversial tactics to reassert control. Stressing the need for "loyalty (…) to the Council,"[92] Chairman Hobhouse and vice-Chairman Burden started to use confidentiality clauses to prevent damaging leaks to the press and discipline unruly Council members.[93] The move angered neutral members. Things came to a head at a Council meeting on April 4, 1973. Following the failure of attempts to expel Seager because too few Council members could attend the meeting,[94] Reform Group members tried to turn the situation into an advantage by calling for a secret vote on condemning fox hunting. Still wielding a relative majority,

[88] RSPCA Archives, CM/61 RSPCA Council Minutes 1972–1975, Meeting of the Council, 06.12.1972, 3–4; the impact of the protest was exacerbated by initial confusion of Bryan Seager's name on the press statement with that of Major Seager, Chief Executive Officer of the RSPCA.

[89] British Library, Richard Ryder Papers, Ryder Dep. 9846 B2;1 RSPCA Reform, 1971–1975, RSPCA Press Release, 21.11.1972.

[90] British Library, Richard Ryder Papers, Ryder Dep. 9846 B2;1 RSPCA Reform, 1971–1975, Press Statement RSPCA Reform Group, 25.11.1972; RSPCA Archives, CM/61 RSPCA Council Minutes 1972–1975; Meeting of the Council, 06.12.1972, 3–7, 11.

[91] RSPCA Archives, CM/60 RSPCA Council Minutes 1971–1972; Meeting of the Council, 06.12.1972, 3–6; Meeting of the Council, 04.04.1973, 4; Ryder, *Animal Revolution*, 173; Robert Garner, *Animals, Politics and Morality* (Manchester and New York: Manchester University Press, 1993), 55–56.

[92] RSPCA Archives, CM/60 RSPCA Council Minutes 1971–1972; Meeting of the Council, 27.07.1972, 2; Meeting of the Council, 06.12.1972, 7.

[93] RSPCA Archives, CM/61 RSPCA Council Minutes 1972–1975; Meeting of the Council, 06.12.1972, 3–7; Meeting of the Council, 07.02.1973, 5–7.

[94] RSPCA Archives, CM/61 RSPCA Council Minutes 1972–1975; Meeting of the Council, 04.04.1973, 4.

Chairman Hobhouse reacted by calling for drastic changes to the RSPCA's constitution. Hobhouse's camp tabled two motions for the Society's next general meeting. The first motion proposed raising the maximum number of non-elected co-opted Council members from six to ten. The second motion proposed a formal rule allowing a three quarter Council majority to eject a member from Council. If implemented, the motions would have had a significant effect on Council dynamics: in 1973, there were 24 elected Council members, 16 voting representatives of associated groups, and between 3 and 6 voting co-opted members. Co-opted members were proposed in Council and elected by a simple majority. Raising their number to ten meant that the majority of voting members on the Council could eventually be non-elected. Meanwhile, the option of expelling elected members from Council had troubling implications for the Society's democratic constitution. Despite fierce protests from attending neutral and Reform Group members, the proposed motions were passed by a 'traditionalist' majority.[95]

The passage of the motions triggered a wider organisational crisis. Within a month of the April 1973 Council meeting, a members' petition with 500 signatures demanded an Extraordinary General Meeting. Although some of the petition's content was deemed defamatory by the Society's legal counsel, Hobhouse was advised to delay tabling the motions and agree to demands for an impartial enquiry of RSPCA leadership and management: "It seems to me that you have nothing to lose."[96]

Established after the Society's annual general meeting in June 1973, the three-man panel of enquiry under lawyer Charles Sparrow, QC, submitted its report in late 1974.[97] The outcome was a blow to Hobhouse. Reviewers made allegations of mismanagement, called for the resignation of Hobhouse, and proposed streamlining the RSPCA's eight standing committees, one ad hoc committee, and three advisory committees, which cost at least £20,000 to service per year.[98] Other proposals included halving the number of Council members, making Council membership depen-

[95] RSPCA Archives, CM/61 RSPCA Council Minutes 1972–1975; Meeting of the Council, 04.04.1973, 4–7; 12–13; Special Meeting of the Council, 03.04.1974, 9–10.

[96] RSPCA Archives, CM/61 RSPCA Council Minutes 1972–1975, Special Meeting of the Council, 24.05.1973, 2.

[97] RSPCA Archives, CM/61 RSPCA Council Minutes 1972–1975, Special Meeting of the Council, 03.04.1974, 1.

[98] RSPCA Archives, CM/61 RSPCA Council Minutes 1972–1975, Statement by Executive Director at Meeting of the Council, 01.11.1973 (item 10(8(c))); Ryder, *Animal Revolution*, 173–175; "RSPCA gets new chairman", *Times*, 16.01.1975, 2.

dent on relevant expertise, reducing a chairperson's term of office to two years, holding a referendum on 'field sports,' and updating voting procedures at general meetings.[99] Hobhouse, who had been re-elected as chairman with a 26 to 7 Council majority in July 1974,[100] felt compelled to offer his resignation in November 1974 and formally resigned in January 1975.[101] Writing to Reform Group member Richard Ryder, he felt that further confidence votes would have deepened Council rifts but complained about the panel of enquiry's "devious methods."[102]

RSPCA reforms were voted on by an extraordinary general meeting in 1975.[103] One year later, a streamlined Council unanimously decided to "oppose all hunting with hounds."[104] The Reform Group dissolved itself in May 1975. Although their methods had been criticised by the 1974 enquiry report, Reform Group members had implemented nearly all of their core demands and had also attained senior positions within the Society.[105] Richard Ryder in particular emerged as an influential figure pushing for a reorientation of RSPCA campaigning. Replacing Hobhouse's vice-chairman Frank Burden in 1976,[106] Ryder acknowledged that the Society's public image had been damaged by leaks, perceptions of a "divided" council and member base, criticism of the Royal Family, and concerns about an excessive focus on "bloodsports (or dogs/cats)."[107] Ryder tried to refocus campaigning on 'positive' policies. Elected as RSPCA Chairman in 1977, these policies would include a new focus on

[99] RSPCA Archives, CM/61 RSPCA Council Minutes 1972–1975, Special Meeting of the Council, 03.04.1974; Roscher, *Königreich*, 294–97, Ryder, *Animal Revolution*, 173–75.

[100] RSPCA Archives, CM/61 RSPCA Council Minutes 1972–1975; Meeting of the Council, 31.07.1974, 1.

[101] RSPCA Archives, CM/61 RSPCA Council Minutes 1972–1975; Meeting of the Council, 21.11.1974, 1; Special Meeting of the Council, 08.–09./ 15.01.1975, 16.

[102] British Library, Richard Ryder Papers, RSPCA Ryder Dep. 9856, B2/2, RSPCA Reform, 1971–1972, 2. 1974–1975, John Hobhouse to Ryder, 20.01.1975; Garner, *Animals, Politics and Morality*, 56–57.

[103] RSPCA Archives, CM/61 RSPCA Council Minutes 1972–1975; Meeting of the Council, 21.11.1974, 1; Special Meeting of the Council, 08.–09./ 15.01.1975, 17–18.

[104] Quoted according to Ryder, *Animal Revolution*, 175.

[105] British Library, Richard Ryder Papers, RSPCA Ryder Dep. 9856, B2/2, RSPCA Reform, 1971–1972, 2. 1974–1975, Stanley Cover, Announcement—RSPCA Reform Group, 26.05.1975.

[106] RSPCA Archives, CM/61 RSPCA Council Minutes 1972–1975, Meeting of the Council, 31.07.1974, 2–3.

[107] British Library, Richard Ryder Papers, Ryder Dep. 9846 B1, 4, Richard Ryder to Mike Seymour-Rouse, 20.03.1976, 1.

animal rights, a more active engagement with other European animal and environmentalist organisations, and large-scale campaigns against seal hunting and Britain's decision to restart live animal exports in 1977.[108] Farm animal welfare was also an important focus. In 1978, the 149th RSPCA general meeting unanimously voted: "That this Society accepts a commitment to making a full scale effort as a priority, to combat the suffering caused to millions of animals in intensive farming systems, experimentation, zoos, circuses, and safari parks and other areas of mass-exploitation."[109]

Losing Touch: Harrison and RSPCA Reform

Ruth Harrison was side-lined by events. Distrusted by RSPCA leadership after leaking the BFSS letter and criticised by younger campaigners for her FAWAC work, she was unable to find allies in either the 'traditionalist' or 'reform' camp.

During parliamentary discussions of revised welfare codes between 1970 and 1971, the formerly close alliance between Harrison, RSPCA leadership, and the Parliamentary Animal Welfare Group was no longer intact.[110] On July 30, 1971, criticism by the RSPCA and Parliamentary Animal Welfare Group made Minister of Agriculture James Prior guarantee that new space standards for calf pens would enable animals to groom themselves and agree to introduce minimum iron levels in calf feeds.[111] Harrison played no role in the campaign. After unsuccessfully proposing an offshoot RSPCA Society for animal welfare in June 1970,[112] she was ousted as chair of the Society's ad hoc farm animals committee. When the ad hoc committee met to discuss FAWAC code reviews in October 1970,

[108] British Library, Richard Ryder Papers, Ryder Dep. 9846, F5/7, Greenpeace and Friends of the Earth, 1979–1987; B1/5 Minutes and memoranda of RSPCA Council meetings, 1971–1978; B3/2 Export of Live Animals, 1973–1978; B3/3 Export of Live Animals, 1978–1989; B1/4 Major R. Seager; M. Seymour-Rouse.

[109] British Library, Richard Ryder Papers, Ryder Dep. 9846, B1/5, Minutes and memoranda of RSPCA Council Meetings, Summary of Proceedings 149th Annual General Meeting, 23.06.1978; Garner, *Animals, Politics and Morality*, 56–57.

[110] RSPCA Archives, CM/59 RSPCA Council Minutes 1970–1971; Meeting of the Council, 26.11.1970, 2; Meeting of the Council, 05.08.1971, 4.

[111] RSPCA Archives, IL/25/1 RSPCA Intensive Farming 2 of 2, The RSPCA and Livestock. Report No. V. 11. Produced by the Veterinary Department of the RSPCA for the Panel of Enquiry. February 1974, 3.

[112] RSPCA Archives, CM/59 RSPCA Council Minutes 1970–1971, Meeting of the Council, 04.06.1970, 3.

John Hobhouse and Frank Burden raised concerns about Harrison's role in reviewing codes proposed by her own committee—something that had not bothered them one year earlier.[113] Harrison had walked out of the meeting but subsequently denied she had resigned.[114]

Things did not improve following the foundation of the RSPCA's new Farm Livestock Advisory Committee (FLAC) in June 1971.[115] With the main Council's attention increasingly occupied by 'field sports' controversies and campaigns against live animal exports,[116] FLAC soon handled nearly all day-to-day farm animal welfare business. Although she was invited to join FLAC in Spring 1972 and participated in debates on new farrowing systems for pigs,[117] Harrison found that her status as 'lay expert' counted little in a committee dominated by welfare scientists, ethologists, professional farmers, and veterinarians (Chap. 10).

Harrison's relationship with the RSPCA Reform Group was equally difficult. This was in part due to her status as a 'traditionalist' in the eyes of influential reformers like Richard Ryder.[118] Crediting the Reform Group with rejuvenating the RSPCA and ending the dominance of older upper- and upper-middle class women,[119] Ryder later remembered

[113] RSPCA Archives, Ad Hoc Committees—1969–1971, Meeting of the ad hoc Farm Animals Committee, 15.10.1970, 1.

[114] RSPCA Archives, Ad Hoc Committees—1969–1971, Meeting of the ad hoc Farm Animals Committee, 15.10.1970, 1; RSPCA Archives, CM/59 RSPCA Council Minutes 1970–1971, Meeting of the Council, 26.11.1970, 5; Meeting of the Council, 05.08.1971, 2.

[115] RSPCA Archives, Minutes of the Farm Livestock Advisory Committee [subsequently FLAC] Meetings held between 20.07.1971 and 20.05.1975, FLAC Meeting, 20.07.1971, 1.

[116] RSPCA Archives, CM/59 RSPCA Council Minutes 1970–1971, Meeting of the Council, 01.04.1971, 1; RSPCA campaigning and BBC publicity had led to a halt of exports in 1973; SELFA campaigning was reactivated when Britain restarted live animal exports after a two-year moratorium in 1975 and confirmed its position during a 1977 enquiry; at the 1978 Secretaries' Conference, motions on the "general welfare of Farm Livestock in Britain" to improve animal housing and improve slaughter techniques were agreed on as a "sequel to the SELFA Campaign"; British Library, Richard Ryder Papers, Ryder Dep. 9846, B1/5, Minutes and memoranda of RSPCA Council Meetings, RSPCA Secretaries' Conference, 22.06.1978, 1; see also SELFA timeline in British Library, Richard Ryder Papers, Ryder Dep. 9846, B3/2, Export Live Animals, 1973–1978, David Wilkins document; B3/3, NFU Pamphlet "no case to answer".

[117] RSPCA Archives, Minutes FLAC, Meeting 24.04.1972, 1 & 5.

[118] Relations with other Reform Group members like John Bryant were better, British Library, Richard Ryder Papers, RSPCA Ryder Dep. 9856, B2/2, RSPCA Reform, 1971–1972, 2. 1974–1975, RSPCA Reform Group News Letter, February 1974—Ruth Harrison, April 1974.

[119] Roscher, *Königreich*, 364–365.

Harrison as "sensitive and sincere but not easy to get on with."[120] Despite acknowledging *Animal Machines*' significance,[121] he saw Harrison as a "rather right wing" campaigner, who feared "opposition from those she saw as being in a position of authority … Government officials, MPs, RSPCA people etc."[122]

As described by Mieke Roscher and Emily Gaarder, this description of Harrison is fairly typical of the gender imagery associated with 'second-wave' 1970s and 1980s animal activism. Since the nineteenth century, female activists had formed the majority of members in most British campaigning organisations. However, gender stereotypes had often led to a marginalisation of female voices and leaders. Whereas activists like Harrison had previously been downplayed as overly emotional 'crazed spinsters' (see Part III), the 1970s and 1980s frequently saw them accused of being too sentimental or timid to take on leadership roles and fight effectively in the rejuvenated 'virile' world of radicalised protest.[123]

Relations between Ryder and Harrison seem to have gotten off to a rocky start. After joining the RSPCA Council in mid-1971, Ryder was approached by Harrison's solicitors in December 1972. He was asked to provide "a written account" of a Council meeting on December 6, "particularly dealing with the way the conduct of the meeting demonstrated the attitudes and pressures brought to bear on particular members of the Council."[124] The report would help "convince the jury" in Harrison's libel suit against Nadia Nerina "of the difficulties under which any minority Council member works who is not willing to fall in with the viewpoint of the hierarchy."[125] Ryder could provide evidence on:

> the manner in which [the Council] suppress[es] the viewpoints wither by threats of expulsion or by in fact expelling members. I fear that in practice one such witch hunt is likely to come to a conclusion in February when the

[120] Correspondence with Richard Ryder (10.08.2015).

[121] Ryder, "Harrison, Ruth (1920–2000)".

[122] Correspondence with Richard Ryder (10.08.2015).

[123] Roscher, *Königreich*, 366, 361–370; see also Gaarder, *Women and the Animal Rights Movement*, 94–116.

[124] British Library, Richard Ryder Papers, Ryder Dep. 9846 B2/1 RSPCA Reform, 1971–1975, Halsey Lightly and Hemsley to Richard Ryder, 14.12.1972.

[125] British Library, Richard Ryder Papers, Ryder Dep. 9846 B2/1 RSPCA Reform, 1971–1975, Halsey Lightly and Hemsley to Richard Ryder, 14.12.1972.

> Society's hierarchy will be represented at the inquisition by a QC. This could well be useful to us in showing how the hierarchy expends the funds available.[126]

Concerned about whether reporting confidential details would leave him on thin legal ice, Ryder refused. When threatened with a subpoena, he, however, agreed to provide vague answers to a list of questions prepared by Harrison's legal team.[127] Soon afterwards, Harrison failed to guarantee her support for Brian Seager ahead of the RSPCA Council's 1973 meeting to decide on whether to expel him.[128]

Disagreement between Harrison and Ryder extended beyond Council politics to whether it was ethically justifiable to campaign for improvements of intensive farming or whether it was necessary to totally oppose the practice. The son of wealthy landowners and a former hunter,[129] Ryder had played a leading role in the contemporary rise of animal rights thinking and was part of the so-called Oxford Group. Consisting of students, researchers, and activists, the Oxford Group loosely came together to discuss the philosophical and ethical dimensions of human–animal relations during the late 1960s. Amongst the Group's members were Stanley and Roslind Godlovitch, John Harris, David Wood, and Michael Peters.[130] According to moral philosopher, Peter Singer, who was also affiliated with the Oxford Group, the impact of Harrison's book on members had been "enormous":[131]

> together with my wife, Renata, I met Richard's wife, Mary, and the two other Canadian philosophy students, Roslind and Stanley Godlovitch, who had been responsible for Richard and Mary becoming vegetarians. … They had come to see our treatment of non-human animals as analogous to the

[126] British Library, Richard Ryder Papers, Ryder Dep. 9846 B2/1 RSPCA Reform, 1971–1975, Halsey Lightly and Hemsley to Richard Ryder, 14.12.1972.

[127] British Library, Richard Ryder Papers, Ryder Dep. 9846 B2/1 RSPCA Reform, 1971–1975, Richard Ryder to Alan A. Meyer, 09.01.1973.

[128] British Library, Richard Ryder Papers, Ryder Dep. 9846 B2/1 RSPCA Reform, 1971–1975, Letter: With the compliments of Ruth Harrison, Attached letter to Major Seager, 06.03.1973.

[129] Roscher, *Königreich*, 296.

[130] Roscher, *Königreich*, 267; Garner and Okuleye, *The Oxford Group and the Emergence of Animal Rights.*

[131] Correspondence with Peter Singer (17.01.2015).

> brutal exploitation of other races by whites in earlier centuries. This analogy they now urged on us, challenging us to find a morally relevant distinction between humans and non-humans which could justify the difference we make in our treatment of those who belong to our own species and those who do not. During these two months, Renata and I read Ruth Harrison's pioneering attack on factory farms, *Animal Machines*.[132]

While Singer was converting to vegetarianism, John Harris and the Godlovitches were preparing an edited volume titled *Animals Men and Morals*.[133] Amongst the contributors to the book's "factual" section were Richard Ryder and Ruth Harrison. In her chapter "On Factory Farming," Harrison mixed earlier *Animal Machines* material with criticism of recent regulatory changes, which promoted the spread of intensive production systems. Harrison also attacked controversial practices like the castration of calves; hatcheries' 'sexing' lines, which discarded unwanted chicks into rubbish bins, where they suffocated; and the chemical caponisation of cocks.[134] Since 1965, the government had missed several opportunities for meaningful welfare reforms: "If the statutory regulations urged by the Brambell Committee had been implemented quickly they would have proved acceptable to farmers in general."[135] Activists and consumers thus faced an ethical dilemma:

> Most people accept the position of eating meat only on condition that the animal has pleasure in life while it lives and is then humanely slaughtered. In no instance can these two criteria be guaranteed today. Many people have become so repulsed by the situation that they have taken the first step towards opting out of it by becoming vegetarians. ... The vegan ... takes the most logical step towards elimination of cruelty, a step to which only a very small but gallant minority have so far devoted their lives.[136]

[132] Peter Singer, "Animal Liberation: A Personal View," *Between the Species*, 2/(3):18 (1986), 149.

[133] Roslind Godlovitch, Stanley Godlovitch, and John Harris (eds.), *Animals, Men and Morals. An Enquiry into the Maltreatment of Non-Humans* (London: Victor Gollancz, 1971).

[134] Ruth Harrison, "On Factory Farming," in Roslind Godlovitch, Stanley Godlovitch, and John Harris (eds.), *Animals, Men and Morals. An Enquiry into the Maltreatment of Non-Humans* (London: Viktor Gollancz Ltd, 1971), 12–17.

[135] Harrison, "On Factory Farming," 19.

[136] Harrison, "On Factory Farming," 23.

However, by itself, gallant consumerism would not transform existing markets. Even if life-long vegetarians like Harrison abstained from meat consumption, the effects on overall demand would be insufficient to make intensive systems unprofitable and end animal suffering. In this situation, opposing animal husbandry per se would do far less for animals than campaigning for an improved "biologically and ethically acceptable"[137] mode of animal production.

Ryder's chapter "Experiments on Animals"[138] was far more radical in its attack on animal exploitation per se. Employing the term *speciesism*, Ryder argued that humans and animals were situated on a moral continuum. If racist discrimination was immoral amongst humans, *speciesist* discrimination against animals by humans for the sake of experiments—and by extension intensive food production—was equally reprehensible. Four years later, Peter Singer provided further intellectual support for per se opposition of experimentation and livestock production. In his 1975 *Animal Liberation*, Singer argued that animal exploitation violated the Benthamite principle of equal consideration of interests by going against animals' interest in not suffering.[139]

Although both acknowledge the importance of *Animal Machines*, neither Singer nor Ryder agreed with Harrison's efforts to reform an intensive agricultural system to which they were opposed in principle. Meeting her "once or twice" during conferences, Singer remembers talking to Harrison "about tactics":

> I thought she was too conservative, in terms of how to go about achieving change, ... she was for slow incremental reform, and had greater hopes for [FAWAC] than I did. I wanted more public campaigning, protests, encouragement of vegetarianism, etc.[140]

When Ryder organised a major 1977 RSPCA symposium on the "Ethical Aspects of Man's Relationship with Animals," Harrison was invited but notably absent from the list of 150 signatories of the resulting declaration

[137] Harrison, "On Factory Farming," 23.

[138] Richard D. Ryder, "Experiments on Animals," in Roslind Godlovitch, Stanley Godlovitch, and John Harris (eds.), *Animals, Men and Morals* (London: Victor Golancz Ltd, 1971); Garner and Okuleye, *The Oxford Group*, 85–99.

[139] Singer, Peter, *Animal Liberation. A New Ethics for Our Treatment of Animals* (New York: Harper Collins, 1975); Singer, "Animal Liberation: A Personal View".

[140] Correspondence with Peter Singer (17.01.2015).

on animal rights and against speciesism.[141] As a committed pacifist, Harrison also rejected the occasionally violent activism of radicalising segments of the animal rights movement.[142] Despite being in contact with members of the Animal Liberation Front (ALF) and sympathising with the emerging animal rights philosophy of thinkers like Tom Regan, Harrison was adamant in her rejection of unethical and unproductive violence.[143] According to animal welfare scientist Donald Broom:

> She knew people in the Animal Liberation Front and she was very careful not to cause them any direct problems like … passing on their names or anything like that. …, but she didn't agree with any violence. … she might have sympathy with what they were trying to achieve, [but] she thought it was the wrong thing to do. And she did feel that some animal research was justified, so she wasn't in favour of the more extreme actions.[144]

Too radical for RSPCA traditionalists, whom she further alienated by publicly attacking the Society's electrothanator for killing stray dogs in 1974,[145] and too moderate for the RSPCA Reform Group and younger activists, Harrison failed to secure re-election in the postal ballot for the streamlined RSPCA Council in mid-1975.[146] Although she remained a member of many animal protection societies ranging from the Animal Defence Society to the Catholic Study Circle for Animal Welfare,[147]

[141] British Library, Richard Ryder Papers, B4/1 Papers relating to the RSPCA Symposium, "The Rights of Animals", 1977, Correspondence and papers 1977–1978, Report of the Subcommittee of the Council to Enquire into the feasibility of holding a two-day symposium on the ethical aspects of man's treatment of animals, 21.01.1977; The Rights of Animals. A Declaration Against Specism.

[142] Keith Tester, "The British Experience of the Militant Opposition to the Agricultural Use of Animals," *Journal of Agricultural Ethics* 2 (1989), 241–251.

[143] Oral History Interview Donald Broom (04.07.2014).

[144] Oral History Interview Donald Broom (04.07.2014).

[145] RSPCA Archives, CM/61 RSPCA Council Minutes 1972–1975, Meeting of the Council, 16.05.1974, 4; Meeting of the Council, 31.07.1974, 10; Meeting of the Council, 13.11.1974, 6; Statement of the Council on the Electrothanator by JM Bryant and Mrs R. Harrison; Meeting of the Council, 02.01.1975, 4–6; Harrison and Byrant's criticism led to design modifications: RSPCA Archives, CM/61 RSPCA Council Minutes 1972–1975, Meeting of the Council, 08.05.1975, 7–8.

[146] RSPCA Archives, IF/56/6, Ruth Harrison General File, RM/A853, Memorandum: Archivist to HFA, 11.12.1993, 1.

[147] RSPCA Archives, IF/56/6, Ruth Harrison General File, RM/A853, Memorandum: Archivist to HFA, 11.12.1993, 1.

Harrison subsequently concentrated on her official FAWAC work and on strengthening her own Farm Animal Care Trust (FACT). From the mid-1970s onwards, FACT would not only fund her own campaigning but also allow her to strengthen relations with the fledgling discipline of animal welfare science.

CHAPTER 10

Slippery FACTs: The Rise of a "mandated" Animal Welfare Science

By 1975, scientific engagement with animal welfare had come a long way since *Animal Machines*. In 1973, ethology's status had been given a significant boost with the award of the Nobel Prize in Physiology or Medicine to Konrad Lorenz, Niko Tinbergen, and Karl von Frisch for "their discoveries concerning organization and elicitation of individual and social behaviour patterns."[1] Meanwhile, earlier taboos about studying animals' affective states were falling as a result of the retirement of first-generation ethologists and a wider reorientation of the discipline. In Britain, younger researchers were beginning to apply behavioural findings to farms, and a new discipline of animal welfare science was emerging out of ethology and the veterinary sciences. From the beginning, this new discipline was a "mandated science."[2] With regulators, activists, and producers all trying to define welfare, researchers were supposed to set 'objective' standards and profited from a resulting increase of funding.[3] Defining standards proved difficult. The Brambell Committee had made important proposals for improving animal welfare. However, there was still much uncertainty about how to define and measure welfare and how to turn resulting

[1] "The Nobel Prize in Physiology or Medicine 1973", *The Nobel Prize*, https://www.nobelprize.org/prizes/medicine/1973/summary/ (20.05.2020).

[2] David Fraser, "Understanding animal welfare," *Acta Veterinaria Scandinavica* 50 Supl (2008), 6.

[3] Robert Kirk, "The Invention of the 'Stressed Animal'," 241–263.

C. Kirchhelle, *Bearing Witness*, Palgrave Studies in the History of Social Movements,
https://doi.org/10.1007/978-3-030-62792-8_10

findings into practical rules: how much stress was acceptable? How should adequate diets be defined? How much water did a cow require? Initial hopes for easily quantifiable welfare markers soon proved misguided. Although researchers continued to emphasise that measuring welfare was possible,[4] resulting methodologies and experimental setups were influenced by disciplinary preferences, contemporary value struggles, and target-oriented funding from sponsors like the RSPCA and Ruth Harrison.

A New Kind of Science: The Rise of Animal Welfare Science

In the decade following the 1965 Brambell Report, three different approaches to assessing welfare emerged: (1) a first approach evaluated classic physiological indicators of animals' basic health and biological functioning, (2) a second approach employed a mix of physiological and behavioural methods to study how animals 'cope' with farm environments; and (3) a third approach focused on the 'naturalness' of different production systems.[5]

Veterinary scientists were among the first to engage with farm animal welfare research. The discipline's engagement is unsurprising given veterinarians' traditional role as guardians of animal health, close ties to the agricultural industry, and parallel calls for preventive veterinary health services in Britain.[6] Their alignment with the productivist paradigm of post-war agriculture guaranteed veterinary researchers a full hearing in ministries and committees like FAWAC. It also made many early veterinary researchers focus less on behavioural and more on quantifiable physiological welfare measures with which to optimise intensive systems.[7] For example, 1960s' research by Swedish veterinarian Ingvar Ekesbo studied blood calcium levels and teat injury frequency to assess different dairy cow systems.[8] Measuring physiological indicators of stress was deemed particularly important. While much of early stress research was later criticised as

[4] Donald M. Boom, "A History of Animal Welfare Science," *Acta Biotheoretica* 59 (2011), 127–137.

[5] Buller and Roe, *Food and Animal Welfare*, 31–35.

[6] Woods, "Is Prevention Better Than Cure?".

[7] Woods, "Cruelty to Welfare," 19–20; Marian [Stamp] Dawkins, *Animal Suffering. The Science of Animal Welfare* (London and New York: Chapman and Hall, 1980), 109–111.

[8] Ingvar Ekesbo, "Disease incidence in tied and loose housed dairy cattle and causes," *Acta Agriculturae Scandinavica* 15 (Suppl) (1966), 1–74.

simplistic,[9] increasingly sophisticated models evolved during the 1970s. For a while, this raised hopes that husbandry systems could be evaluated by precise measurements of cytokine, corticoid, and other hormone levels in animals' blood. Health was thought to be the opposite of stress, whose ultimate manifestations were disease and death.[10] According to veterinary researcher and FAWAC member David Sainsbury, "good health" was the "birthright" of every animal: "If it becomes diseased we have failed in our duty to the animal and subjected it to a degree of suffering that cannot be readily estimated."[11]

While early veterinary welfare research provided important insights for the design of husbandry systems, its alleged overemphasis on physiological markers was criticised by other disciplines. The early 1970s saw a growing number of ethologists follow the example of W. H. Thorpe and study farm animal behaviour as an indicator of welfare.[12] This engagement was in part a continuation of post-war shifts towards acknowledging animals' affective states and in part a response to new funding opportunities.[13] It also accelerated the fragmentation of classic ethology.

Now described as a "lost discipline,"[14] classic ethology reached the pinnacle of its international visibility in the 1960s and early 1970s. Despite ongoing resistance by Lorenz, the late 1950s had seen earlier 'hydrological' models of behaviour abandoned in favour of development- and ecology-oriented models. In 1963, Tinbergen made his last major theoretical contribution to ethology with his four whys (function, phylogeny, mechanism, and ontogeny) to animal behaviour.[15] Around the world, there were now dedicated ethological journals, conferences, and centres of learning. In public, prize-winning documentaries by Tinbergen and the

[9] David Fraser, "Biology of Animal Stress. Implications for Animal Well-being," *Journal of Applied Animal Welfare Science* 2/2 (1999), 157–159.

[10] David Fraser, *Understanding Animal Welfare. The Science in Its Cultural Context* (Oxford: UFAW, 2008), 122–123; Robert Kirk, "Invention of the 'Stressed Animal'," 255–258.

[11] David Sainsbury, *Farm Animal Welfare. Cattle, Pigs and Poultry* (London: Collins, 1986); David Fraser, "Understanding Animal Welfare," *Acta Veterinaria Scandinavica* 50 (supplement 1), S 2.

[12] Boom, "A History of Animal Welfare Science," 124.

[13] Kirk, "The Invention of the 'Stressed Animal'".

[14] "Ethology: Claims and Limits of a lost Discipline", Podcast Series Wissenschafts Portal Gerda Henkel Stiftung, https://lisa.gerda-henkel-stiftung.de/ethology_claims_and_limits_of_a_lost_discipline_podcast_series?nav_id=9149 [15.05.2021].

[15] Burkhardt, *Patterns of Behavior*, 418–420, 427–434.

TV shows by zoologist Desmond Morris helped popularise the new discipline for mass-audiences.[16]

However, ethology's rising star also unleashed centrifugal forces. Among senior ethologists, a major point of contention centred on whether ethological animal behaviour models could be applied to other contexts. While Tinbergen remained concerned about diluting ethology's authority by overstating findings, Konrad Lorenz extended already outdated hydrological models to humans in his 1963 book *On Aggression*. According to Lorenz, aggression was an evolutionary force in human history that could trigger catastrophes but could also be controlled by collective releases like Olympic Games.[17] The book proved controversial. Some senior ethologists like William Thorpe modified Lorenz's claims but defended the underlying notion that ethological explanations of behaviour could drive the future development of social sciences like sociology and psychology.[18] By contrast, long-standing critics of ethology like UK Chief Scientific Adviser, Solly Zuckerman, used Lorenz's book to attack the discipline for overselling its findings.[19] In 1967, Oxford zoologist Desmond Morris further destabilised early ethology's focus on animal behaviour. In the *Naked Ape*, Morris applied ethological approaches to human evolution and sexual selection—including endorsements of being naked and much-criticised claims about alleged female traits and social roles.[20] Tinbergen disassociated himself from such claims. In his 1968 inaugural lecture as Oxford's Professor of Animal Behaviour, he noted that ethology could help explain human behaviour but that existing claims were "no more than likely guesses."[21]

While *On Aggression*, *Naked Ape*, and the 1973 Nobel Prize focused public attention on the work of older ethologists, younger researchers were already breaking new ground. In contrast to classic ethology's focus on causation, studies of neurophysiology, sociobiology, and evolutionary

[16] Burkhardt., *Patterns of Behavior*, 440, 444–446; Hans Kruuk, *Niko's Nature. The Life of Niko Tinbergen and his Science of Animal Behaviour* (Oxford: Oxford University Press, 2003), 221; Morris hosted *Zootime* between 1956 and 1967.

[17] Konrad Lorenz, *Das Sogenannte Böse. Zur Naturgeschichte der Aggression* (Wien: Borotha Shoeler, 1963).

[18] W.H. Thorpe, "Zoology and Behavioural Sciences," *Nature* 216/5110 (07.10.1967), 20.

[19] Solly Zuckerman, "The Human beast," *Nature* 212/5062 (05.11.1966), 563–564; see also: "Is Ethology Respectable?," *Nature* 216/5110 (07.10.1967), 10.

[20] Desmond Morris, *The Naked Ape. A Zoologist's Study of the Human Animal* (New York: McGraw Hill, 1967).

[21] Quoted according to Burkhardt, *Patterns of Behavior*, 440.

function became the "rage"[22] in animal behavioural sciences after William Hamilton's path-breaking 1964 publications on kin selection and E.O. Wilson's *New Synthesis* in 1975. Rather than querying behavioural motivation at the level of the individual, younger researchers used quantitative socio-biological approaches to study how traits and 'selfish genes' structured behaviour at the group level. Others focused on the adaptive dimensions of animal behaviour in an ecological context or on behaviour's cognitive and affective dimensions.[23]

For an increasing number of researchers in this latter group, animal welfare offered a comparatively well-funded niche to expand into and satisfy growing calls for ethology to become more 'applied.'[24] Their growing involvement with farm animal welfare brought applied ethologists into contact and conflict with veterinary researchers prioritising physiological measurements of stress. Writing to *Nature* in 1969, William Thorpe had already criticised the fact that the government's FAWAC "included no member capable of expressing authoritatively the current views and results of scientists working on animal behaviour."[25] The dominance of non-ethological voices on FAWAC had led to a disregard of Brambell recommendations and a narrow equation of welfare with the absence of pain. Their lack of engagement with affective states and behaviour meant that resulting FAWAC codes were "a retrograde step and would, in fact, condone in farming practice operations which would not be tolerated (…) in a scientific or medical laboratory in this country without special license."[26]

However, ideas about how exactly ethological approaches could be applied to farm animal welfare varied. Although there was agreement that behaviour mattered, there was no consensus on what a behaviour meant and how it could be correlated with physiological data.[27] The result was a prolonged phase of what science and technology studies scholars describe

[22] Burkhardt, *Patterns of Behavior*, 464.

[23] Burkhardt, *Patterns of Behavior*, 460–465.

[24] Burkhardt, *Patterns of Behavior*, 458; Robert Kirk, "The Invention of the 'Stressed Animal'," 253–259.

[25] W.H. Thorpe, "Welfare of Domestic Animals," *Nature* 224/5214 (04.10.1969), 18; for the wider evolution of contemporary notions of stress see Robert Kirk, "The Invention of the 'Stressed Animal'," 241–263.

[26] W.H. Thorpe, "Welfare of Domestic Animals," 20.

[27] David Fraser, *Understanding Animal Welfare*; Robert Kirk, "The Invention of the 'Stressed Animal'," 253–259.

as scientific "tinkering"[28] with different groups testing distinct methodological approaches and trialling creative experimental setups.

Some researchers tried to assess whether farm animals' behaviour could be classified as normal or abnormal when compared to a species' 'natural' behaviour in the 'wild.'[29] This approach proved especially popular among veterinary researchers. Despite many veterinarians' focus on stress (see above), ethology was by no means an alien field within the veterinary sciences. Founded in 1966, the Society for Veterinary Ethology—now the International Society for Applied Ethology—was initially open only to veterinarians and drew on two centres of learning in Cambridge and Edinburgh.[30]

In Cambridge, veterinarian Alastair Worden's Institute for the Study of Animal Behaviour (ISAB) had recruited veterinarians to study 'wild' animals since 1944 and began to extend its focus to domestic animals in the late 1950s.[31] Another centre of veterinary ethology emerged in Edinburgh. In 1959, Scottish veterinarian Andrew Fraser conducted important ethological studies on 'abnormal' behaviour in farm settings. Fraser was subsequently encouraged to promote animal ethology at Edinburgh's Royal (Dick) College of Veterinary Medicine and drew on local support from a group of researchers working at Edinburgh's Poultry Research Centre (now Roslin Institute).[32]

Headed by animal geneticist David Wood-Gush, the Poultry Research Centre group compared animal behaviour in 'natural' and agricultural

[28] Karin D. Knorr, "Tinkering toward Success: Prelude to a Theory of Scientific Practice," *Theory and Society* 8/3 (1979), 347–376; Helen Curry, *Evolution Made to Order: Plant Breeding and Technological Innovation in Twentieth-Century America* (Chicago: University of Chicago Press, 2016), 9; Dmitriy Myelnikov, "Tinkering with genes and embryos: the multiple invention of transgenic mice c. 1980," *History and Technology* 35/4 (2019), 425–452.

[29] [Stamp] Dawkins, *Animal Suffering*, 39–40; 110–111.

[30] Andrew Fraser, "A Short Biography of Andrew Fraser, written by him in March 2008," *Applied Ethology*, https://www.applied-ethology.org/res/dr_%20andrew%20fraser_%20isae%20honorary%20fellow.pdf [11.03.2020].

[31] J.C. Petherick and Ian J.H. Duncan, "The International Society for Applied Ethology: going strong 50 years on," in Jennifer Brown, Yolande Seddon and Michael Appleby (eds.), *Animals and Us – 50 years and more of applied ethology* (Wageningen: Wageningen Academic Publishers, 2016), 30–31.

[32] Andrew Fraser, "Displacement activities in domestic animals," *British Veterinary Journal* Vol. 115 (1959), 195–200; David Fraser, *Understanding Animal Welfare. The Science in its Cultural Context* (Oxford: UFAW, 2008), 125; Petherick and Duncan, "The International Society for Applied Ethology," 30.

settings.[33] In an influential 1973 review, Wood-Gush challenged the Brambell Report's equation of welfare with the freedom to express 'natural' behaviour. It was possible to breed birds for better adaptation to battery cages, cage designs did not necessarily frustrate so-called displacement or vacuum behaviours, and not every behavioural frustration was detrimental to welfare. This did not mean that affective states were irrelevant. Comparative research on wild fowls and battery hens indicated that social isolation and lack of stimulus increased the likelihood of feather pecking or cannibalism. However, to be of practical relevance, welfare observations had to be reconciled with economic realities. Conventional farrowing systems' frustration of nest building by sows could be characterised as a welfare problem even as they prevented the crushing of piglets.[34] Because the avoidance of all behavioural frustration on farms was "impossible,"[35] researchers should focus on understanding which frustration was tolerable and which was damaging. Wood-Gush's group subsequently began to experiment with the rewilding of farm animals and alternative rearing systems. This included releasing intensively farmed pigs into the semi-wild environment of the Edinburgh Pig Park and using results to design a new family pen housing system for farms.[36]

While Wood-Gush employed a comparative approach to define 'normal behaviour' and welfare, other researchers focused on evolutionary adaptiveness to reconcile ethology with existing notions of thrift. Having trained under Thorpe before becoming the first person to hold a chair in animal welfare in 1986, Reading- and Cambridge-based biologist Donald M. Broom proposed that welfare be used only to refer to an animal's state and not to external benefits given to it. According to Broom, welfare could be assessed on a scale running from good to bad and was directly related to an animal's ability to cope with its immediate environment. Coping could be measured via classic physiological indicators like survival

[33] Newberry and Sandilands, "Pioneers of applied ethology," 57–58.

[34] D.G.M. Wood-Gush, "Animal Welfare in Modern Agriculture," *British Veterinary Journal* 129 (1973), 167–173.

[35] D.G.M. Wood-Gush, "Animal Welfare in Modern Agriculture," *British Veterinary Journal* 129 (1973), 173.

[36] Alex Stolba & D.G.M. Wood-Gush, "The identification of behavioural key features and their incorporation into a housing design for pigs," *Annal Recher Vét* 15 (1984), 287–298; Victoria Sandilands, "David Wood-Gush The Biography of an Ethology Mentor," *Applied Animal Behaviour Science* 87 (2004), 173–176; David Fraser, *Understanding Animal Welfare*, 169–176.

into old age and the ability to produce offspring. Stress in this context remained a negative concept, which overtaxed an individual's control systems and reduced fitness. Affective states were important but were only one component contributing to an animal's welfare.[37]

For other researchers, understanding what animals wanted was key to defining welfare. While Wood-Gush and Broom did not deny that affective states were important, they were not particularly interested in using them as a door through which to comprehend an animal's point of view. Establishing this point of view became the focus of an influential third group of mostly ethology-trained researchers. Since the 1960s, an increasing amount of research on animal consciousness, such as preference indication by rats and mirror self-awareness among chimpanzees, had overthrown many earlier taboos about studying animals' mental states.[38] In 1976, US ethologist Donald Griffin's influential *The Question of Animal Awareness* expanded earlier claims by researchers like William Thorpe and argued that non-humans had consciousness: "it even seems likely that we can anticipate the eventual emergence of a truly experimental science dealing with the mental experiences of other species."[39]

In Oxford, Tinbergen student Marian Stamp Dawkins built on Griffin's work to propose a new preference-oriented approach to welfare. Dawkins cautioned against defining welfare solely in terms of stress and (ab)normal behaviour or by trying to understand animal preferences via intuition.[40] Instead, she designed experiments that queried animals' preferences and how they adapted to different environments.[41] In 1977, she tested whether hens preferred battery or larger pens. To her surprise, the animals spent equal amounts of time in both settings.[42] Responding to criticism that this

[37] Broom, "A History of Animal Welfare Science," 128–133; Fraser, *Understanding Animal Welfare*, 72–73.

[38] [Stamp] Dawkins, *Animal Suffering*, 16–26; Katja Guenther, "Monkeys, Mirrors, and Me: Gordon Gallup and the Study of Self-Recognition," *Journal of the History of the Behavioural Sciences* 53/1 (2017), 5–27.

[39] Donald Griffin, *The Question of Animal Awareness. Evolutionary Continuity of Mental Experience* (New York: Rockefeller University Press, 1976), 14.

[40] [Stamp] Dawkins, *Animal Suffering*, 1–54; Marian Stamp Dawkins, "From an animal's point of view: Motivation, fitness, and animal welfare," *Behavioural and Brain Sciences* 13 (1990), 1–3.

[41] [Stamp] Dawkins, *Animal Suffering*, 110–116.

[42] Marian [Stamp] Dawkins, "Do hens suffer in battery cages? Environmental preferences and welfare," *Animal Behaviour* 25 (1977), 1034–1046; Fraser, *Understanding Animal Welfare*, 191–193.

result might merely show what animals were accustomed to,[43] she next fused (neo-)liberal economic and ethological theories by designing choice experiments in which animals worked towards certain goals. The harder an animal was willing to work for something, the more important this factor seemed to be for its welfare. Results were then calibrated against work performed to achieve commonly agreed-on welfare components like food. The 'price' an animal was willing to pay for different factors could be traded and compared. 'Price elasticity' could be measured by varying required workloads and demand could be assessed by minimising 'income' in the form of limiting time to access a resource.[44]

For Dawkins and other members of the 'feelings school,' including Wood-Gush student Ian Duncan, welfare definitions had to encompass an animal's mental and physical health, harmony with its environment or the ability to adapt without suffering to a new artificial environment, and a consideration of an animal's feelings and preferences. Suffering was not just constituted by pain and stress but by a wide range of unpleasant emotional states including subjective feelings like boredom.[45]

Set out in her influential 1980 book *Animal Suffering: The Science of Animal Welfare*,[46] Dawkins' preference-centred research approach bore an uncanny resemblance to contemporary economic and political theories emphasising individual needs and desires.[47] It is a long-standing truism in the history of science that science is always also a reflection of the society in which it is conducted. This is particularly true for animal welfare science. As described by animal scientist David Fraser, its focus on providing scientific parameters for inherently normative concepts meant that animal welfare research was both "science-based" and "values-based."[48]

[43] Ian J. H. Duncan, "The interpretation of preference tests in animal behaviour," *Applied Animal Ethology* 4 (1978), 197–200.

[44] [Stamp] Dawkins, *Animal Suffering*, 83–97; Fraser, *Understanding Animal Welfare*, 73, 197–203.

[45] Ian J.H. Duncan and Marian Stamp Dawkins, "The problem of assessing 'well-being' and 'suffering' in farm animals," in D. Smidt et al. (eds.) *Indicators relevant to farm animal welfare* (The Hague: Martinus Nijhoff, 1983), 13–24.

[46] [Stamp] Dawkins, *Animal Suffering*.

[47] In 1990, Dawkins contextualised her application of economic theory by noting that other biological fields and psychologists had also adopted economics of choice to explain population dynamics and behaviour; Stamp Dawkins, "From an animal's point of view," 4–5.

[48] Fraser, *Understanding Animal Welfare*, 273.

In laboratories, in official circles, and on farms, researchers constantly triangulated between three different needs: the need to design experimental systems that generated 'objective' and reproducible welfare data, the need to propose parameters that were meaningful to the producers and politicians who would have to adopt them, and the need to align research programmes with wider societal notions of welfare. Emerging metrics and standards thus not only reflected researchers' individual methodologies but also represented an often-unstable compromise between what different academic disciplines, non-academic sponsors, agricultural producers, and consumers valued most.

No part of this hybrid research and standard-setting process was easy. There was a basic consensus that welfare could be measured. There was also an increasing amount of disciplinary coordination with the already mentioned Society for Veterinary Ethology gaining more members and launching a dedicated journal (*Applied Animal Ethology*, now *Applied Animal Behaviour Science*) in 1974.[49] However, despite many methodological overlaps,[50] the three academic schools of fitness, affective states, and 'natural' rearing continued to disagree on the relative weighting and usefulness of different welfare indicators.[51]

Combative Science: Welfare Research in a Time of Counter Science

All the while, pressure to apply welfare research was increasing. With more British animals being produced by fewer workers,[52] scientists were asked to evaluate existing systems and to define welfare standards for the design of new intensive and non-intensive systems. For researchers, this was both a chance and a challenge. In 1972, British government policy shifted from supporting comparatively independent research councils that sponsored 'basic research' towards supporting government departments that sponsored 'applied research.' As described by Dmitriy Myelnikov, this shift was

[49] Suzanne Millman, Ian Duncan, Markus Stauffacher, and Joseph Stookey, "The impact of applied ethologists and the International Society for Applied Ethology in improving animal welfare," *Applied Animal Behaviour Science*, 86 (2004), 299–311; Andrew Fraser, "A Short Biography of Andrew Fraser".

[50] Broom, "A History of Animal Welfare Science," 127.

[51] Fraser, *Understanding Animal Welfare*, 83, 230–232.

[52] Yago Zayed, "Agriculture: historical statistics," *House of Commons Library* Briefing Paper 03339 (2016), 7–10.

particularly pronounced with regard to the Agricultural Research Council, which saw about half of its budget moved to MAFF.[53] Their focus on applied research meant that welfare researchers stood to profit from this development. Enterprising behavioural researchers could also tap into a growing number of non-governmental funding streams—something that would become increasingly important in the uncertain 1980s funding environment for universities (Chap. 12).[54]

The increase of funding sources required shrewd diplomacy on the part of scientists.[55] To maintain their public standing as trustworthy experts, welfare researchers had to not only minimise internal disagreements on methodologies but also find ways of managing funder expectations without compromising research integrity. Appearing 'objective' and resisting pressure to produce favourable results was particularly important during a time of widespread polarisation over what constituted acceptable welfare and whose expertise to trust. Similar to parallel controversies about environmental expertise, the 1970s saw formerly 'backstage' decision-making about animal welfare give way to increasingly antagonistic 'frontstage' public clashes between officials, producers, and activists over whose version of welfare to trust (Chap. 8).[56] As records from Ruth Harrison's FACT and the RSPCA show, being able to commission one's own (counter) science and mobilise experts in committees, parliament, and public became key campaigning assets—albeit ones that caused problems with more radical campaigners, who opposed all forms of animal experimentation.[57]

In the case of the RSPCA, the 1970s saw the Society tread new ground by actively courting animal welfare researchers (Chap. 9). Significantly, the Society's historical opposition to 'cruel' animal experimentation made it focus on commissioning behavioural rather than physiological research. The RSPCA's 1969 general meeting had passed a resolution originally

[53] Dmitriy Myelnikov, "Cuts and the cutting edge: British science funding and the making of animal biotechnology in 1980s Edinburgh," *The British Journal for the History of Science* 50/4 (2017), 708–709; boundaries between both categories of science were ideal type and blurred, there was also no pronounced antagonism between the ARC and MAFF.

[54] Jon Agar, *Science Policy Under Thatcher* (London: UCL Press, 2019), 21–22, 73–86, 263–265.

[55] For developments in the US see: *Agricultural Research: Background and Issues* (Washington DC: Congressional Research Service, 2020), 12–14.

[56] Matthias Heymann, "1970s: Turn of an Era in the History of Science?," *Centaurus* 59.1–2 (2017), 1–9; Hilton et al., *Politics of Expertise*, 14–15; Cassidy, *Vermin*, 48–49, for front- and backstage dimensions of policy see 205.

[57] Harrison and the RSPCA were not unique in bringing "expertise to expertise", Hilton et al. *Politics of Expertise*, 81–82.

tabled by Ruth Harrison according to which the Society should spend "more of its resources and influence on the welfare of farm animals."[58] The Society's leadership had responded by hiring an additional veterinary officer in 1970 to give welfare advice and by remodelling its existing farming committee along "rather different lines": "The new committee would be composed essentially of certain Council members of the RSPCA, together with some acknowledged experts in the veterinary, farming and ethological fields from outside the Council of the RSPCA."[59]

After a "considerable search,"[60] a number of renowned experts agreed to help. At its first meeting on June 7, 1971, the RSPCA's new Farm Livestock Advisory Committee (FLAC) consisted of a zoologist, an ethologist, veterinarians, full-time farmers, a representative of the British Horse Society, and a lawyer specialising in animal jurisprudence.[61] Maintaining their scientific independence was very important to FLAC experts. According to University of London primatologist and FLAC chairman John R. Napier, FLAC "was an independent committee and to have any credibility in the eyes of the general public, it must be seen to be objective and uninfluenced."[62] In contrast to some RSPCA activists, FLAC was not opposed to intensive farming per se. Its mission was rather "to define what was reasonable in animal welfare terms"[63] and press for regulatory improvements. According to Napier, "we had, as a society, undoubtedly to accept that domestication was inevitable, but the degree to which this was taken must be closely watched. We had to draw the line between comfort and

[58] RSPCA Archives, IF/25/1 RSPCA Intensive Farming 2 of 2, The RSPCA and Farm Livestock. Report No. V. 11 produced by the Veterinary Department of the RSPCA for the Panel of Enquiry. February 1974, 5.

[59] RSPCA Archives, IF/25/1 RSPCA Intensive Farming 2 of 2, The RSPCA and Farm Livestock. Report No. V. 11 produced by the Veterinary Department of the RSPCA for the Panel of Enquiry. February 1974, 5.

[60] RSPCA Archives, IF/25/1 RSPCA Intensive Farming 2 of 2, The RSPCA and Farm Livestock. Report No. V. 11 produced by the Veterinary Department of the RSPCA for the Panel of Enquiry. February 1974, 6.

[61] Anon, "Society for Veterinary Ethology, 'Stress in Farm Animals' – proceedings of joint symposium with the Royal Society for the Prevention of Cruelty to Animals, London 25–26, May 1973," *British Veterinary Journal* 130 (1974), 85; the Society's veterinary department under Chief Veterinary Officer Philip Brown was closely affiliated with FLAC.

[62] RSPCA Archives, Minutes FLAC, Meeting, 12.10.1971, 3.

[63] RSPCA Archives, IF/25/1 RSPCA Intensive Farming 2 of 2, The RSPCA and Farm Livestock. Report No. V. 11 produced by the Veterinary Department of the RSPCA for the Panel of Enquiry. February 1974, 7.

discomfort and cruelty, and time and again we were brought back to the conclusion that the animals' behaviour may be our best barometer."[64]

FLAC's first job was to advise the RSPCA with regard to the revised 1971 draft animal welfare codes (Chap. 8). Although it commended new veterinary rights of entry, it noted that MAFF's State Veterinary Service was understaffed and not carrying out welfare inspections. FLAC was even more critical of FAWAC, which had failed "to reform and educate in animal welfare. No significant research had been sponsored, nor the right questions asked, nor the relevant new information publicised."[65] In the case of welfare codes for veal husbandry, FLAC pushed for systems that provided sufficient space for calves to groom themselves, optimum rather than minimum dietary iron levels, and roughage to aid rumen development. It was hoped that consumer information would prevent intensive veal systems from further spreading in the UK.[66]

Between 1972 and 1976, the number of items on FLAC's agenda grew rapidly. Responding to requests by the RSPCA Council, MAFF, and members of the public, the committee reviewed the non-stun slaughter of animals, early fowl weaning, the introduction of new battery rearing systems for piglets, a 'protecta' farrowing system to protect piglets from being crushed by sows, castration, the docking of pigs' tails, and live animal exports.[67]

Rather than consulting experts on an ad hoc basis, FLAC also pushed for long-term RSPCA sponsorship of the new "sub-discipline of farm livestock ethology."[68] In October 1971, FLAC chairman John Napier noted that FLAC should:

> be forward looking and that it would require ethological evidence if it were to make valid recommendations. Very little work had to date been done on

[64] RSPCA Archives, Minutes FLAC, Meeting, 09.02.1972, 2.

[65] RSPCA Archives, IF?25/1 RSPCA Intensive Farming 2 of 2, The RSPCA and Farm Livestock. Report No. V. 11 produced by the Veterinary Department of the RSPCA for the Panel of Enquiry. February 1974, 5.

[66] RSPCA Archives, Minutes FLAC, Meeting, 20.07.1971, 3; Meeting, 12.10.1971, 2–3.

[67] RSPCA Archives, Minutes FLAC, Meeting, 09.02.1972, 1–2; Meeting, 24.04.1972, 2–3, 4–5; Meeting, 27.02.1973, 1–2; Meeting, 14.05.1973, 3; Meeting, 04.12.1973, 2; Meeting, 09.09.1975, 2, 4; Meeting, 27.04.1976, 4; Meeting, 01.09.1976, 2–3; RSPCA Archives, Minutes FLAC, Report on Intensive Farming Produced for RSPCA Council by the Veterinary Department, 1–2.

[68] Anon, "Society for Veterinary Ethology, 'Stress in Farm Animals' – proceedings of joint symposium," 85.

> animal behaviour under intensive farming conditions. We needed more information, and one way of obtaining it would be to sponsor certain selected lines of work in the ethological field.[69]

According to Napier, producing "ethograms" or "repertoires of the nature and context of every classifiable action, fixed motor patterns, performed by farm animals"[70] was more than a disinterested scientific endeavour. It would help the Society "interpret the meaning of 'unnecessary suffering', the key phrase in the 1911 *Protection of Animals Act* and all subsequent legislation"[71] and aid the development of regulations for abnormal stress. Although RSPCA Vice-Chairman Frank Burden insisted that "we did not need scientific evidence that certain practices were cruel, or wrong, we knew they were,"[72] Napier's initiative was supported by other FLAC members. According to veterinary ethologist Andrew Fraser, sponsoring research would be an "excellent public relations exercise": "there were large areas where descriptive observations would be of much value to [FLAC]."[73] Quoting Liverpool animal researcher Roger Ewbank, Fraser claimed that "one of the first signs (and possibly the only sign) of stress/ distress/ deprivation of intensively kept farm animals may be the development of abnormal behaviour patterns."[74] The attending BVA president Nigel Snodgrass agreed that "this information was desperately needed."[75] Following extensive debate, FLAC decided to call for welfare research scholarships, which would not cost much and "ensure that the reins were firmly"[76] in the RSPCA's hands.

RSPCA sponsorship of behavioural research was approved by the Society's Council in early 1972 on the condition that research would be observational and "non-experimental."[77] This limitation to non-

[69] RSPCA Archives, Minutes FLAC, Meeting, 12.10.1971, 4.

[70] Anon, "Society for Veterinary Ethology, 'Stress in Farm Animals' – proceedings of joint symposium," 85.

[71] Anon, "Society for Veterinary Ethology, 'Stress in Farm Animals' – proceedings of joint symposium," 85.

[72] RSPCA Archives, Minutes FLAC, Meeting, 12.10.1971, 4.

[73] RSPCA Archives, Minutes FLAC, Meeting, 12.10.1971, 4; Fraser resigned from FLAC due to other commitments in 1974; RSPCA Archives, Minutes FLAC, Meeting, 25.06.1974, 1.

[74] RSPCA Archives, Minutes FLAC, Meeting, 09.02.1972, 3.

[75] RSPCA Archives, Minutes FLAC, Meeting, 09.02.1972, 3.

[76] RSPCA Archives, Minutes FLAC, Meeting, 09.02.1972, 3.

[77] RSPCA Archives, Minutes FLAC, Meeting, 24.04.1972, 2.

experimental research had important consequences: it effectively excluded research on physiological indicators of abnormal stress and pain from RSPCA sponsorship and simultaneously generated important pump-priming funds for the budding community of applied ethologists.[78]

The first project financed by FLAC was a comprehensive survey of behavioural welfare research in the UK. Between July and November 1972, one of Napier's graduate students, Peter Lattin,[79] visited 21 academic institutions and 50 researchers across the UK. In his report, Lattin noted that while many people were interested in behavioural work, "one very often felt it was a political and economic interest."[80] To avoid sponsoring biased research, FLAC should fund theoretical work by trained ethologists in zoology departments and field work in agricultural and veterinary colleges.[81] Only very few research centres had the capacity to serve as bases for useful applied research.

Building on Lattin's report, FLAC awarded the first RSPCA Farm Animal Behaviour Awards in May 1973. The two groups chosen were based at the University of Sussex and the Edinburgh School of Agriculture.[82] At Sussex, Dr Marthe Kiley and Prof Richard Andrew received £2000 to conduct behavioural research on calves and mother/calf relationships between birth and weaning in intensive and extensive veal units.[83] In Edinburgh, David Wood-Gush's group used RSPCA sponsorship to study the excretory behaviour of pigs in extensive systems to better understand behavioural problems in farrowing crates and sow stalls. RSPCA funding was later also used to study the effects of providing bedding material during farrowing and to co-finance Edinburgh's semi-wild herd of intensively bred pigs.[84]

[78] Anon., "Society for Veterinary Ethology, 'Stress in Farm Animals' – proceedings of joint symposium," 86.

[79] RSPCA Archives, Minutes FLAC, Meeting, 18.09.1972, 1.

[80] RSPCA Archives, Minutes FLAC, Meeting, 14.11.1972, 2.

[81] Anon., "Society for Veterinary Ethology, 'Stress in Farm Animals' – proceedings of joint symposium," 86.

[82] RSPCA Archives, Minutes FLAC, Meeting, 14.05.1973, 2.

[83] RSPCA Archives, Minutes FLAC, Meeting, 25.06.1974, 5; RSPCA Archives, IL/24/5 RSPCA Intensive Farming 1 of 2, RSPCA Grant Annual Report 1973/1974 RJ Andrew and M. Kiley; RSPCA Archives, IF/25/1 RSPCA Intensive Farming 2 of 2, The RSPCA and Farm Livestock. Report No. V. 11 produced by the Veterinary Department of the RSPCA for the Panel of Enquiry. February 1974, 9.

[84] RSPCA Archives, Minutes FLAC; Meeting, 23.11.1978, 4; Meeting, 01.05.1979, 2; RSPCA Archives, IL/24/5 RSPCA Intensive Farming 1 of 2, John Napier, Farm Animal

FLAC lobbying also led to RSPCA conference sponsorship. In May 1973, the RSPCA co-organised the Society for Veterinary Ethology's 1973 symposium on *Stress in Farm Animals* at the London Zoo.[85] This was the first ever "purely scientific"[86] meeting organised by the Society since its foundation in 1824. Covered by the *New Scientist*,[87] conference proceedings were dominated by physiological research on stress but also featured important behavioural contributions by Andrew Fraser, Roger Ewbank, and Ian Duncan.[88] During the opening ceremony, John Napier praised RSPCA chairman John Hobhouse's decision to move the RSPCA "towards a more scientifically oriented approach to the problems of animal welfare."[89] FLAC grants would hopefully help animal behaviour become a "meeting ground for psychologists, zoologists, physiologists, anatomists, geneticists, ecologists, and many others."[90]

Napier's vision for FLAC sponsorship of interdisciplinary behavioural research was ultimately constrained by the RSPCA Council's limitation of Behaviour Awards to non-experimental research. Where exactly the line between observational and experimental research lay was unclear. Attempts to define it not only exposed differences of opinion between FLAC researchers and their RSPCA sponsors but also highlighted the limits of expert power in the value-laden field of welfare science. Historically, the experiment had functioned as a carefully staged way to produce 'self-evident' truths and trust in the authority of learned experts.[91] Traditional

Behaviour Awards, 1974–1975; Interim Report Edinburgh University; FLAC later also considered financing limited studies on the economic feasibility of different systems for laying hens, RSPCA Archives, FLAC Minutes, Meeting, 23.11.1978, 1; Inlay: The Proposed Straw Yard Project and Alternative Systems".

85 RSPCA Archives, Minutes FLAC, Meeting, 18.09.1972, 2.

86 RSPCA Archives, IF/25/1 RSPCA Intensive Farming 2 of 2, The RSPCA and Farm Livestock, Report No. V. 11 Produced by the Veterinary Department of the RSPCA for the Panel of Enquiry. February 1974, 18.

87 "Domesticated animals breed on regardless," *New Scientist*, 14.06.1973, 665.

88 Anon., "Society for Veterinary Ethology, 'Stress in Farm Animals' – proceedings of joint symposium," 85–95.

89 Anon., "Society for Veterinary Ethology, 'Stress in Farm Animals' – proceedings of joint symposium," 85.

90 Anon., "Society for Veterinary Ethology, 'Stress in Farm Animals' – proceedings of joint symposium," 86.

91 A concept most famously set out in Steven Shapin and Simon Schaffer, *Leviathan and the air-pump: Hobbes, Boyle, and the experimental life* (Princeton: Princeton University Press, [1985] 2011).

expert authority was, however, inherently limited in mandated sciences like animal welfare science, which simultaneously tackled moral and scientific claims. While older experts like Thorpe and Huxley had developed scientific principles of humane experimentation as a way to re-establish science's progressivist moral role after the Second World War (Chap. 4), younger activists challenged scientists' authority to define ethically acceptable experimentation.

Tensions about what constituted moral research and who was allowed to define it escalated when FLAC decided to sponsor behavioural research at the University of Oxford. After turning down earlier proposals,[92] FLAC had awarded £2000 in 1974 to Marian Dawkins to study poultry behaviour. Dawkins wanted to assess animals' welfare preferences by rearing and then splitting a group of egg-laying hens in two. To avoid confirming learnt preferences, one group of birds would become accustomed to an extensive free range system and the other would be held in standard intensive conditions, with two birds sharing a cage of 15 inches width. Intensive birds would then be allocated to the extensive free range system and vice versa and given an option to choose the system they preferred.[93] Although Dawkins' research did not require a Home Office License for animal experimentation and involved no manipulation of animals' bodies, the idea of rearing animals in battery cages led to tensions between scientific and non-scientific members of FLAC and the RSPCA Council. While researchers like Peter Lattin, John Napier, and zoologist Peter Jewell commended Dawkins' research as "the most interesting project that had been put before the Committee,"[94] lay members including Ruth Harrison and Richard Ryder condemned it as a "cruel experiment."[95]

The conflict about what constituted an 'experiment' drew in the wider RSPCA Council and highlighted underlying tensions about whether one should sponsor animal research to improve welfare or oppose all forms of animal exploitation in principle. Following the resignation of RSPCA Chairman Hobhouse (Chap. 9), a joint meeting between FLAC and the RSPCA's Animal Experimentation Advisory Committee on September 2,

[92] RSPCA Archives, Minutes FLAC, Meeting, 25.06.1974, 6.

[93] RSPCA Archives, IL/24/5 RSPCA Intensive Farming 1 of 2, M. Dawkins and D. McFarland, Welfare of Domestic Fowl in Relation to their habitat preferences.

[94] RSPCA Archives, Minutes FLAC, Meeting, 18.03.1975, 3.

[95] RSPCA Archives, IL/24/5 RSPCA Intensive Farming 1 of 2, Letter Major Seager, 12.08.1975.

1975, led to an éclat. At the meeting, Dawkins' research was attacked by Richard Ryder, Labour peer Lord Houghton of Sowerby, and parapsychologist and Doctor Who adviser Christopher "Kit" Pedler, who claimed that the Oxford experiments were unscientific. According to Ryder, the RSPCA was "on terribly dangerous ground here": "We have been giving away money in a rather unbusinesslike way, trusting people with whom we are scarcely acquainted, working in laboratories where only occasional inspections have been made by the Society. (…). Obviously the RSPCA must never inadvertently subsidise cruelty."[96]

The RSPCA Council subsequently declined an extension of the behavioural awards. This decision enraged FLAC scientists. Resigning as FLAC chairman seven days after the September éclat, John Napier expressed pride in his committee's achievements. Within four years, FLAC had become recognised "as an important body" of welfare expertise, had published the first RSPCA-commissioned research, was organising its second research conference on mutilation, and had pushed the RSPCA to "take part in the academic side of animal welfare."[97] According to Napier, "it was wrong to describe [Dawkins'] work as experimentation (…) it was very difficult to interpret ethological studies if one did not have any knowledge of, or sympathy with the subject of ethology."[98] The fact that the proposed research did not require a Home Office license had been confirmed by the Home Office, Dawkins, and Oxford's Head of Animal Behaviour Research. Attending RSPCA Council members disagreed: "the Council had been preoccupied during the last two years and probably not made an adequate study of the documents sent to them by the FLAC. Being now reduced in size the new Council was anxious to do everything possible to repair the damage done to the Society by the events of the preceding two years."[99] From now on, decisions over experiments' acceptability would not be made by welfare scientists but by the Council.

Acknowledging its defeat, FLAC decided to no longer refer to behavioural studies "using the terms 'research' or 'experimentation' as these

[96] British Library, Richard Ryder Papers, Ryder Dep. 9846 B1, 4, Richard Ryder to Major R. Seager, Executive Director, RSPCA, 18.09.1975, 2–3.

[97] RSPCA Archives, FLAC Minutes, Meeting, 09.09.1975, 1.

[98] RSPCA Archives, FLAC Minutes, Meeting, 09.09.1975, 4; see also: RSPCA Archives, IL:24:5 RSPCA Intensive Farming 1 of 2, Jon Napier: Farm Animal Behaviour Awards, 1974–1975.

[99] RSPCA Archives, FLAC Minutes; Meeting, 09.09.1975, 5.

could confuse, and possibly worry, certain members of the Council."[100] Although the Council resumed payment of the awards, rewording research briefs failed to solve problems. In November 1975, FLAC formulated detailed terms and conditions and reporting requirements for RSPCA funding.[101] This triggered further tensions with the Council, which had followed suggestions by Ryder and stipulated unannounced control visits by lay Council members.[102] Warning that universities would resist such a move,[103] FLAC instead proposed visits by the Society's Chief Veterinary Officer or a nominated Council representative. FLAC's new head Prof. Peter Jewell of London's Royal Holloway College hoped that such a person would ideally be "suitably qualified" with a "perhaps easier understanding of the studies taking place, and less possibility of an emotional or arbitrary attitude."[104] FLAC also pressed for one of its members to be there for any visit.[105]

FLAC eventually got its way on research visits and also managed to secure a time-limited funding extension for Dawkins, who was summoned twice to describe her research.[106] However, the committee's overall influence within the RSPCA was waning. FLAC recommendations restructured the Society's farm animal welfare policies in 1977[107] and members influenced new European welfare codes.[108] But scientists' attempts to push for a middle ground between "commercial farming interests" and "over-emotional and misleading"[109] attacks on intensification fell on deaf ears. Both the RSPCA Council and general meetings increasingly condemned intensive farming per se. In the case of live animal exports, FLAC failed to prevent RSPCA calls for a complete ban, as opposed to banning only the export of animals

[100] RSPCA Archives, FLAC Minutes, Meeting, 09.09.1975, 5.

[101] RSPCA Archives, FLAC Minutes, Meeting, 11.11.1975, 3.

[102] British Library, Richard Ryder Papers, Ryder Dep. 9846 B1, 4, Richard Ryder to Major R. Seager, Executive Director, RSPCA, 18.09.1975, 2.

[103] RSPCA Archives, FLAC Minutes, Meeting, 10.02.1976, 2–3.

[104] RSPCA Archives, FLAC Minutes, Meeting, 27.04.1976, 3

[105] RSPCA Archives, FLAC Minutes, Meeting, 09.03.1977, 4.

[106] RSPCA Archives, FLAC Minutes, Meeting, 19.10.1976, 1; 24.01.1978, 5.

[107] RSPCA Archives, FLAC Minutes, Meeting, 06.09.1977, 3; Report on Intensive Farming Produced for RSPCA Council by the Veterinary Department, 3–11.

[108] RSPCA Archives, FLAC Minutes, Meeting, 20.05.1975, 1; 09.09.1975, 8; 20.07.1976, 3; 01.05.1979, 4.

[109] RSPCA Archives, FLAC Minutes, Report on Intensive Farming Produced for RSPCA Council by the Veterinary Department, 2–3; see also disagreements with RSPCA leadership and other welfare organisations about condemning intensification during the Animal Welfare Year, RSPCA Archives, FLAC Minutes, Meeting, 12.01.1977, 2.

destined for slaughter.[110] There were also tensions about whether FLAC could participate in meetings with the NFU, BVA, and MAFF, or whether this would undermine negotiations on animal exports being carried out by the Council in tandem with other welfare organisations.[111]

Increasingly critical of its independent expert committee, the RSPCA Council refused to support a proposed RSPCA Chair or centre for animal welfare in May 1977.[112] It also informed FLAC that it would no longer support research on animals in unnatural environments. Referring to Dawkins' work, the Council wanted to make it "quite clear to all concerned that a study of animals in an existing environment was quite in order, but putting animals into a specific environment, which perhaps was not in line with the Society's policies, would not be acceptable."[113] Dismayed FLAC members thought "that it was tragic to have this work stopped just when such excellent results (from the RSPCA's point of view) were beginning to come through."[114] Already struggling to find replacements for the numerous researchers resigning from its ranks,[115] FLAC issued a unanimous rebuttal to accusations that Dawkins' work constituted "bad science": "we considered the results do show how animals may, themselves, indicate their preferences and degrees of comfort, (…). It is much to the credit of the RSPCA to have promoted this work and we wish to reassert our confidence in the value that behavioural studies have in the advancement of animal welfare."[116] Although a site visit by the new RSPCA Chairman Richard Ryder secured a final limited renewal of Dawkins' grant,[117] growing distrust between the Council and FLAC led to a dissolution of the unruly committee and its replacement with a new RSPCA Farming Department in 1979.[118]

The RSPCA was not the only non-governmental organisation to actively engage with and sponsor farm animal welfare science during the

[110] RSPCA Archives, FLAC Minutes; Meeting, 18.09.1972, 3; Meeting, 19.11.1974, 2; Meeting, 21.01.1975, 2; Meeting, 10.02.1976, 5; Meeting, 06.09.1977, 2.

[111] RSPCA Archives, FLAC Minutes; Meeting, 23.05.1978, 4; Meeting, 29.06.1978, 2; British Library, Richard Ryder Papers, Ryder Dep. 9846 B1/4, Major R. Seager to Richard Ryder, 05.08.1977, 1; Ryder to R. Corbett, 09.08.1977, 1; Seager to Ryder, 12.08.1977; Ryder to B. MacDonald, 25.10.1977, 1–2.

[112] RSPCA Archives, FLAC Minutes; Meeting, 11.05.1977, 2.

[113] RSPCA Archives, FLAC Minutes; Meeting, 11.05.1977, 4.

[114] RSPCA Archives, FLAC Minutes; Meeting, 24.01.1978, 5.

[115] RSPCA Archives, FLAC Minutes; Meeting, 23.11.1978, 3.

[116] RSPCA Archives, FLAC Minutes; Meeting, 24.01.1978, 6.

[117] RSPCA Archives, FLAC Minutes; Meeting, 29.06.1978, 1–2; Meeting, 28.09.1978, 1.

[118] RSPCA Archives, FLAC Minutes; Meeting, 01.05.1979, 5.

1970s. Other major organisations like the UFAW began to publish reviews and handbooks on farm animal welfare.[119] Engagement was not limited to large charities. Ruth Harrison also systematically strengthened ties with leading welfare scientists. Her success in doing so was partially due to her involvement in official welfare committees and partially due to the expanded activities of her research trust.

During the early 1970s, the Ruth Harrison Research Trust had developed rather slowly. Trustees met irregularly, and funding decisions were mostly influenced by the ad hoc demands of Ruth Harrison's FAWAC work.[120] The Trust's income also remained based on feet of clay. In 1970, Ruth Harrison appealed to the Marquis of St Innocent "for financial help in work which I feel to be of utmost importance and urgency in helping the animals."[121] While it seemed "unforgiveable to ask for help as soon as I met you," she would "hate to see [the chance to help food animals] slip by" and mentioned that the Trust would "need some £6,000 to take work firmly forward."[122] To raise money, Ruth Harrison also considered co-writing a book on farm systems with Cambridge veterinary scientist David Sainsbury, with whom she enjoyed a good relationship outside FAWAC (Chap. 8).[123] However, no book materialised, and resources remained stretched. Writing to the Whitley Animal Protection Trust in 1971, Harrison stressed that she "desperately" needed "money for general expenses"[124] such as a typewriter, a camera, filing cabinets, and a copying

[119] UFAW, *The UFAW Handbook on the Care and Management of Farm Animals* (Edinburgh: Churchill Livingston, 1971); UFAW continues to sponsor farm animal behaviour and welfare research—the handbook is now in its 5th edition.

[120] FACT Files, DB, Trustees Meetings, Ruth Harrison Research Trust, Meeting of Trustees (13.02.1971).

[121] FACT Files, DB, Unmarked Green Ryman Folder, Ruth Harrison to Marquis of St. Innocent (19.05.1970), 1.

[122] FACT Files, DB, Unmarked Green Ryman Folder, Ruth Harrison to Marquis of St Innocent (19.05.1970), 1; while the Marquis does not seem to have donated any money, he forwarded Harrison's request to 'Woody' [Allen?]; FACT Files, DB, Unmarked Green Ryman Folder, Marquis of St. Innocent to 'Woody' (17.06.1970).

[123] FACT Files, DB, Unmarked Green Ryman Folder, Ruth Harrison to WA Burns (20.09.1971); FACT Files, DB, Dr, Sainsbury, David Sainsbury to Ruth Harrison (30.06.1970); David Sainsbury to Ruth Harrison (08.07.1970); FACT Files, DB, Unmarked Red Ryman Folder Ruth Harrison to David Sainsbury (19.09.1971).

[124] FACT Files, DB, Unmarked Green Ryman Folder, Ruth Harrison to WA Burns (20.09.1971).

machine. In 1973, the Trust's net assets amounted to £4497, while its income fluctuated between £712 (1972) and £959.07 (1973).[125]

It was only following Harrison's exit from the RSPCA Council in 1975 that her Trust's resources and activities experienced a marked increase. FACT financed Harrison's journeys to meetings of the European Standing Committee on animal welfare and networking trips to conferences and animal behaviourists.[126] Harrison soon became a fixture in seminars and conferences on animal welfare. She also joined relevant societies like the Society for Veterinary Ethology/International Society for Applied Ethology (SVE/ISAE).[127] Researcher David Fraser later recalled his first encounter with Harrison at David Wood-Gush's Edinburgh office in the 1970s:

> There was room for two visitors to sit and talk, but any third person had to sit on a chair behind a filing cabinet, hidden from half the room Ruth Harrison took great interest in our research on farm animal welfare and on one of her visits she was occupying that chair … After some time, a colleague … appeared at the door …
>
> 'Ah, I see you've got rid of the good lady,' said the colleague.
>
> 'No,' came the voice of Ruth Harrison from behind the filing cabinet, 'I'm still here.'[128]

In addition to building personal ties to leading academics, Harrison also used FACT to fund welfare research. Much of this research was observational and mirrored RSPCA Council guidance for FLAC. From 1977 onwards, FACT provided valuable pump-priming funding for Prof Anthony (John) Webster's evaluation of different veal production systems by supporting doctoral research on the behaviour, performance, and

[125] FACT Files, DB, Ruth Harrison Research Trust Accounts 30.06.1973; Harrison Research Trust Accounts 30.06.1974.

[126] FACT Files, DB, Unmarked Green Ryman Folder, David Wood-Gush to Ruth Harrison [ca. 1980]; FACT Files, DB, Bristol Veal—Correspondence, Ruth Harrison to Christine (26.06.1979); FACT Files, DB, Unmarked Green Ryman Folder, T.I. Hughes to Ruth Harrison (15.02.1985).

[127] Donald M. Broom, "World Impact of ISAE: past and future," in Jennifer Brown, Yolande Seddon and Michael Appleby (eds), *Animals and Us – 50 years and more of applied ethology* (Wageningen: Wageningen Academic Publishers, 2016), 272.

[128] Fraser, "Ruth Harrison – a Tribute," 17.

health of veal calves in strawyards and crates.[129] FACT also financed physiological research on the stunning of poultry.[130]

Harrison was not engaging science purely for science's sake. Never wavering from the central demands of her 1964 book and learning from her experiences within FAWAC, she saw research as key to supporting her campaign for legislative reform. Her determination to produce useful results made her a hard taskmaster. A critical and prolific reader of scientific publications, Ruth Harrison demanded frequent and detailed progress reports on FACT-financed projects and did not shy away from challenging individual aspects of research.[131] In the case of the Bristol veal studies, this habit caused tensions between Harrison and John Webster. Writing to an acquaintance in 1979, Ruth Harrison complained about Webster's student: "[The student] started work in October last year. … It was on my last visit to the project that I began to have serious doubts about [the student's] real capability of doing the task."[132] Criticising the study's research premises, Harrison also complained that the student had not "read all the papers I had given the department at the beginning of the project."[133] Although she had agreed not to interfere with or bias the research, Harrison mused, "It could be that we will have to insist on changing our post-graduate student."[134]

Three months later, Ruth Harrison reminded Webster that FACT had invested over £5000 in his student's project. Although FACT was grateful for slight amendments to the project, "you had actually promised us … a list of the specific results which would accrue from our continual funding of the project. …—and frankly, your note did not offer us any assurance that what we were seeking … will actually be achieved."[135] Ruth Harrison

[129] The project had initially been rejected by the ARC as "premature"; Webster, "Tribute to an Inspirational Friend," 7–8; FACT Files, MD, FACT Files, FACT Publications & Publicity Material, Pamphlet FACT, 1; FACT Files, DB, Bristol Veal—Correspondence, FACT, Alternative Systems for Veal Calves. Work in progress or proposed for 1980/1981.

[130] FACT Files, DB, Trustees Meetings, FACT. Report to the Trustees (30.06.1977).

[131] FACT Files, DB, Bristol Veal Correspondence, Ruth Harrison to Professor Webster (18.01.1978).

[132] FACT Files, DB, Bristol Veal—Correspondence, Ruth Harrison to Christine (26.06.1979); despite Harrison's initial doubts, the student went on to have a successful research career and relations between Harrison and the student soon improved; FACT Files, DB, Unmarked Green Ryman Folder, Student to Ruth Harrison (03.01.1989).

[133] FACT Files, DB, Bristol Veal—Correspondence, Ruth Harrison to Christine (26.06.1979).

[134] FACT Files, DB, Bristol Veal—Correspondence, Ruth Harrison to Christine (26.06.1979).

[135] FACT Files, DB, Bristol Veal—Correspondence, Harrison to Webster (24.09.1979), 1.

then listed a number of 'facts' she would like to see confirmed and stressed the need for "positive results at the end of the project."[136] Repeating her complaints about Webster's PhD student, Harrison ended by demanding research documentation personally signed by Webster: "You know that we are keen to continue financing this work, but only if it is going to be productive."[137]

Webster's reply was swift and brusque:

> I realise that £6,000 is a substantial sum for [FACT] and I am very grateful for it. However, you and your sponsors must realise that it constitutes only two year's salary. …, and no new graduate, however brilliant and energetic, is going to (a) solve the problems of rearing veal calves in straw yards, and (b) sell the idea to the industry on their own in two years.[138]

According to Webster, FACT was contributing about 4 per cent of costs to the mainly MAFF- and research council-funded research. Meanwhile, ca. 30–40 per cent of the project's outcomes directly addressed FACT concerns: "Expressed crudely, one might say that it is money well spent."[139] It was also wrong to demand too much of a PhD student, who was only one year into the project: "If [the student] is to attempt too much, in an effort to offer most of the glory to FACT, [the student] will fail because there is too much for one person to do, … [and] also fail to set up a proper platform for the next stage of the work."[140]

At present, everything was going well, and the student was capable of effective animal observation: "Whether this leads [the student] to conclude what you or I may wish [the student] to conclude is impossible to tell until the observations have been analysed correctly."[141] Webster also saw "no point in submitting yet another statement of what is going on

[136] FACT Files, DB; Bristol Veal—Correspondence, Harrison to Webster (24.09.1979), 2.

[137] FACT Files, DB; Bristol Veal—Correspondence, Harrison to Webster (24.09.1979), 3.

[138] FACT Files, DB, Bristol Veal—Correspondence, Webster to Ruth Harrison (15.10.1979), 1.

[139] FACT Files, DB, Bristol Veal—Correspondence, Webster to Ruth Harrison (15.10.1979), 1.

[140] FACT Files, DB, Bristol Veal—Correspondence, Webster to Ruth Harrison (15.10.1979), 1.

[141] FACT Files, DB, Bristol Veal—Correspondence, Webster to Ruth Harrison (15.10.1979), 2.

because the project is now clearly defined and you have the details."[142] Following further demands from Harrison, Webster drew a firm line:

> Surely it is only fair to use your words, that when a Trust agrees to fund a research programme intended to lead to a PhD, that body should at least allow the graduate student and … academic supervisor to direct the work over the normal period of three years in the way that best fits the scientific questions. … I am very grateful to FACT for their support which was particularly welcome as it came at an early stage of development … However, I cannot and will not allow FACT to dictate the research that goes on in my department.[143]

While the Bristol veal project was successfully concluded with continued FACT support, Harrison's conflicts with Webster reveal the difficulties scientists faced in managing funder expectations, as well as the increasing importance of generating targeted welfare data for campaigners like Harrison.

In FAWAC, Harrison urgently needed Webster's data to counter industry research and break a regulatory deadlock over low-level iron feeds. In 1976, FAWAC's research sub-committee had reviewed iron-deficient feeds in intensive veal husbandry. Citing *Animal Machines*, the sub-committee had initially recommended a new minimum iron level in calf feeds of 9g/100ml, and the main FAWAC had compromised on a minimum level of 50mg/kg in June 1976.[144] However, in 1977, the research sub-committee was forced to revise its opinion after new industry data showed no evidence that anaemia was a general problem of intensive veal husbandry.[145] Because advice on minimum iron levels "was not soundly based in scientific findings,"[146] the sub-committee concluded that it could not be passed on to Ministers.

[142] FACT Files, DB, Bristol Veal—Correspondence, Webster to Ruth Harrison (15.10.1979), 2.

[143] FACT Files, DB, Bristol Veal—Correspondence, Andrew Webster to Ruth Harrison (30.09.1980); Webster himself gives a positive account of his collaboration with FACT; Webster, "Ruth Harrison – Tribute to an Inspirational Friend," 7–8.

[144] TNA MAF 369/217 FAWAC, Minutes of 15th Meeting (22.06.1976), 4–5.

[145] TNA MAF 369/217 Iron in Calf Diets. Report by the Research Sub-committee on an appraisal of the scientific basis for the draft Advice to Ministers (AWC/76/26, enclosed in: A. Foreman to FAWAC (06.05.1977), 1–2.

[146] TNA MAF 369/217 FAWAC, Minutes of 16th Meeting (18.05.1977), 11.

Harrison was convinced of the opposite but could no longer base her campaigning on the often-anecdotal evidence and synthesist moral argumentation that had characterised her own book and much of Edwardian and early post-war activism. In the combative context of 1970s' expert politics, she needed data and scientific support to counter industry-friendly codes. In FAWAC, she therefore sought to undermine Meat Research Institute data on haemoglobin levels in calves. This led to prolonged clashes with senior FAWAC physiologists over feeds' actual iron content, the validity of industry studies, and whether anaemia could cause suffering in calves.[147] Unable to generate a FAWAC majority, Harrison resorted to the already familiar tactic of blocking a 'bad' compromise until new data being produced by Webster and others supported her position. In May 1977, David Sainsbury complained to FAWAC chair Prof. Harrison about Ruth Harrison's behaviour at a recent meeting of the General Purposes Sub-committee:

> The position is that as a Subcommittee we had agreed all the proposals when we last met ... In fact I was faced, as you know, with a substantial number of proposals almost entirely from Mrs. Harrison, some of a fundamental nature, which were placed in front of me at the meeting and most of which strangely she had not chosen to even mention at previous G.P. Sub-committee meetings. ... I am afraid Mrs. Harrison's tactics promise to delay progress by means which I consider rather dubious.[148]

Others also complained about pre-arranged bullying tactics by Harrison and her supporters, which undermined FAWAC members' ability to present evidence without being "condemned by 'debating points' and in some cases by sheer slander."[149] Harrison's mixed tactics of political and scientific campaigning nonetheless proved successful. In February 1979, FAWAC's General Purposes Sub-Committee published a list of expected data, which would inform the still-undecided calf code revisions. The list included the FACT-sponsored Bristol calf experiment, whose results would eventually change UK and European legislation.[150]

[147] TNA MAF 369/217 FAWAC, Minutes of 16th Meeting (18.05.1977), 12.

[148] TNA MAF 369/222 DWB Sainsbury to Professor RJ Harrison (24.05.1977); FAWAC, General Purposes Sub-Committee, Extracts from Minutes of 7th Meeting (19.04.1977).

[149] TNA MAF 369/222 FR Bell to RJ Harrison (20.05.1977).

[150] TNA MAF 369/240 FAWAC 18th Meeting (08.02.1979), 1–2; Webster, "Ruth Harrison—Tribute to an inspirational friend," 8.

By this time, MAFF was facing significant pressure to address FAWAC's poor welfare record.[151] Divisions within the committee meant that little progress had been made since the Agriculture (Miscellaneous Provisions) Act of 1968. In the case of turkeys, FAWAC had failed to produce new codes despite eight years of review.[152] MAFF's attempt to create an industry-friendly clearing house for welfare codes had clearly failed (Chap. 8). Powerless, unpopular, and divided, FAWAC was replaced with a reformatted Farm Animal Welfare Council (FAWC) in 1979.[153]

The end of FAWAC and the RSPCA's parallel dissolution of FLAC made 1979 a turning point for British farm animal welfare. Beginning in the 1960s and gathering power in the 1970s, centrifugal forces had dissolved established welfare decision-making structures. No part of Britain's animal welfare scene remained untouched: set up as a corporatist committee, MAFF's FAWAC failed to produce viable compromises between increasingly polarised welfarist and industry positions. The failure of corporatist politics was mirrored in the world of welfare campaigning. In the case of the RSPCA, demands for grassroots democracy and conflicts about elite hunting pastimes coincided with the Society's transformation from a conservative charity into a combative and professional NGO with sophisticated in-house scientific expertise. Reaching the height of their influence around 1977, Reform Group members not only remodelled RSPCA leadership and management but also triggered systemic debates about animal rights and the morality of animal exploitation.

Resulting conflicts about what constituted legitimate research and whether scientists alone could define it were indicative of the challenges faced by the new "mandated"[154] discipline of animal welfare science. Tasked with providing objective evidence for normative claims, welfare researchers profited from increased funding but faced pressure to produce useful results. Different sponsors favoured different methodologies. While industry and government bodies emphasised physiological research on stress and pain, the small but growing community of applied ethologists profited from welfare organisations' funding taboos and growing interest in 'normal' behaviour, 'natural' habitats, and animal preferences.

[151] TNA MAF 369/240 FAWAC, Minutes of 17th Meeting (27.01.1978), 3; FACT Files, DB, Unmarked Blue Ryman Folder, Ruth Harrison—FACT—Opinion (19.04.1978).

[152] TNA MAF 369/272 House of Commons Agriculture Committee—Replies to Questions Enclosed with Dr Jack's Letter of 03.02.1981 to Mr Shillito, 3–4.

[153] Ryder, *Animal Revolution*, 184.

[154] David Fraser, "Understanding Animal Welfare," S1.

However, disagreement on how to weight indicators dampened hopes for universal welfare standards. Fifteen years after *Animal Machines*' rallying cry for farm animal welfare, nobody seemed to be able to agree on how new standards should look and who had the authority to define them.

For Harrison, her 12 years within FAWAC and 6 years within the RSPCA Council had proven to be a mixed blessing. On the one hand, membership in both bodies had guaranteed her political influence even after memories of her bestseller began to fade. On the other hand, her work in a notoriously indecisive government committee exposed her to criticism from younger Reform Group members, while her leaking of the BFSS letter left her isolated among RSPCA traditionalists. After failing to secure a re-election to the RSPCA Council in 1975 and losing influence on the public 'frontstage' of activist politics, Harrison refocused on her 'backstage' government work and intensified her ties to animal welfare scientists. Her strategy of commissioning targeted research via FACT, and delaying FAWAC decisions helped prevent the enactment of weak welfare codes. However, without coordinated 'frontstage' pressure tactics like the ones she had employed around 1970, it also failed to produce meaningful new national codes. Estranged from the new RSPCA Council, isolated in an increasingly divided FAWAC, and forced to file personal bankruptcy in 1975, the second half of the 1970s marked an ebb in Harrison's activist career. For a while, it seemed as though there was no longer any place for the almost 60-year-old in the brave new world of animal politics she had helped to unleash.

PART V

From Éclat to Consensus (1979–2000)

Part V examines the maturation of British farm animal welfare politics, activism, and science from 1979 to Harrison's death in 2000. In 1979, the polarisation of British farm animal welfare debates peaked. The Thatcher government ended a decade of relative neglect of farm animal welfare in Downing Street by replacing FAWAC with a new Farm Animal Welfare Committee (FAWC). The move was an acknowledgement of the growing political importance of animal welfare and a blow to post-war corporatist decision-making in British agriculture. In contrast to its predecessor, FAWC was no longer directly controlled by MAFF, comprised more academics, and explicitly acknowledged the Brambell Committee's five freedoms. The new committee was, however, viewed with suspicion by the Royal Society for the Prevention of Cruelty to Animals (RSPCA). Having just dissolved its own Farm Livestock Advisory Committee, the Society refused to send representatives to FAWC. The protest was short-lived and triggered a membership revolt against Richard Ryder. With anti-FAWC protest failing and British agriculture experiencing a prolonged economic crisis, the 1980s and 1990s instead saw the fulfilment of key demands from *Animal Machines*. Now in her 60s, the book's author achieved recognition as a publicly esteemed welfare campaigner. Animal welfare science also matured. Following its rapid expansion in the 1970s, the discipline gained chairs at major British universities. New assurance schemes by major retailers and welfare charities created additional revenue streams and increasingly displaced traditional government regulation with private standard setting. However, the growth of prestige, standards, and resources did not lead to a synthesis of welfare definitions. Although Harrison's views had become part of mainstream politics by the time of her death in 2000, what animal welfare was, how it could be measured, and where the discipline should go next remained open questions.

CHAPTER 11

From Protest to 'Holy Writ': The Mainstreaming of Welfare Politics

The 1980s brought significant changes to the fortunes of Ruth Harrison and farm animal welfare. Having successfully challenged post-war decision-making in British politics and the RSPCA, the wave of radical animal activism began to lose force. Although a radical fringe continued to bomb farms and animal laboratories, more moderate figures began to wield greater influence in animal politics and campaigning. Associated with neither radicals nor 'traditionalists,' Harrison used her excellent connections to animal welfare researchers to secure a place on the newly founded FAWC in 1979. Over the next two decades, FAWC recommendations played a crucial role in implementing key demands of *Animal Machines*. The book's increasingly visionary status and the enactment of new welfare standards helped Harrison secure prestigious markers of public esteem like an Order of the British Empire (OBE) in 1986, as well as more funding for FACT research.

Harrison's transformation into an establishment figure would have been difficult to predict in 1979. In that year, the dissolution of FAWAC could have ended her access to Whitehall decision-making. Despite her uncomfortable role in FAWAC (see Chaps. 8 and 10), Harrison was surprisingly nominated for the new FAWC.[1] FAWC was officially announced by the Thatcher Administration in December 1979 and was meant to

[1] The nomination to a new committee also enabled Harrison to circumvent maximum term limits for advisory committee members.

C. Kirchhelle, *Bearing Witness*, Palgrave Studies in the History of Social Movements,
https://doi.org/10.1007/978-3-030-62792-8_11

correct the flaws of its divided and ineffective predecessor. Ministers insisted that FAWC "advise as speedily as possible on revisions to the Welfare Codes of Cattle, Pigs, Domestic Fowls and Turkeys."[2] Other FAWC briefs included producing codes that were more detailed and placed "more emphasis on behavioural needs."[3] In a significant move, the government announced that FAWC was "free to publicise its views"[4] without prior MAFF consent and could make non-binding recommendations for welfare improvements. It was also made explicit that welfare consisted of more than the absence of cruelty. Ahead of FAWC's first meeting, welfare scientist John Webster had successfully pushed for an updated version of the 'five freedoms' to be integrated into FAWC's mission statement.[5] Whereas FAWAC conflicts had often centred on the validity of basic welfare criteria, FAWC explicitly stated that welfare codes should provide animals with:

1. Freedom from thirst, hunger, or malnutrition;
2. Appropriate comfort and shelter;
3. Prevention, or rapid diagnosis and treatment, of injury and disease;
4. Freedom to display most normal patterns of behaviour;
5. Freedom from fear.[6]

[2] TNA Webarchives, Farm Animal Welfare Council, Press Statement (05.12.1979), http://webarchive.nationalarchives.gov.uk/20121007104210/http://www.fawc.org.uk/pdf/fivefreedoms1979.pdf [19.12.2014], 1.

[3] TNA Webarchives, Farm Animal Welfare Council, Press Statement (05.12.1979), http://webarchive.nationalarchives.gov.uk/20121007104210/http://www.fawc.org.uk/pdf/fivefreedoms1979.pdf [19.12.2014], 1.

[4] TNA Webarchives, Farm Animal Welfare Council, Press Statement (05.12.1979), http://webarchive.nationalarchives.gov.uk/20121007104210/http://www.fawc.org.uk/pdf/fivefreedoms1979.pdf [19.12.2014], 2.

[5] John Webster, *Animal Welfare. A Cool Eye Towards Eden. A constructive approach to the problem of man's dominion over the animals* (Oxford: et al.: Blackwell Science, [1995] 2007), 11; "2013 Prof John Webster and Prof Peter Sandøe", *UFAW Medal for Outstanding Contributions to Animal Welfare Science*, https://www.ufaw.org.uk/ufaw-medal-for-outstanding-contributions-to-animal-welfare-science/ufaw-medal-for-outstanding-contributions-to-animal-welfare-science-past-awards#webster [10.04.2020].

[6] TNA Webarchives, Farm Animal Welfare Council, Press Statement (05.12.1979), http://webarchive.nationalarchives.gov.uk/20121007104210/http://www.fawc.org.uk/pdf/fivefreedoms1979.pdf [19.12.2014], 1; Buller and Roe, *Food and Animal Welfare*, 31–32; definitions were updated in 1993 to read: (1) Freedom from thirst, hunger, and malnutrition—by ready access to fresh water and a diet to maintain full health and vigour; (2) Freedom from discomfort—by providing a suitable environment including shelter and a

Reversing a long policy tradition and acknowledging ethical considerations as a value per se, MAFF also announced: 'animal welfare raises certain points of ethics which are themselves beyond scientific investigation.'[7]

The appointment of FAWC sounded like the fulfilment of an *Animal Machines* wish list and was both a recognition of its predecessor's failings and a concession by the new Conservative administration to growing public support for farm animal welfare. The move also ended a decade of relative neglect of farm animal welfare in No. 10 Downing Street. As described in Part IV, it had long been comparatively easy in government circles for producers and MAFF officials to side-line welfare demands by Ruth Harrison and others as anthropomorphic and misguided (Chap. 8). With notable exceptions like protests against live animal exports, economic instability and frequent 1970s' changes of government had led to a relative neglect of farm animal welfare in Westminster.[8] This relative disinterest and the resulting 'backstage' stasis of British farm animal welfare politics had suited producer interests and MAFF officials. However, lack of progress had also contributed to a growing polarisation of 'frontstage' welfare debates in the public sphere. Public polarisation and international pressure resulting from Britain's decision to join the EEC and European Council welfare conventions steadily increased pressure on senior politicians to address the stasis of welfare reforms and refrain from unconditional defences of intensification.

Although Margaret Thatcher weakened a more explicit earlier draft, the 1979 Conservative election manifesto claimed that the party would ban certain live animal exports, support EEC reform proposals for animal transportation, and shared popular concerns about welfare: "We shall update the Brambell Report, the codes of welfare for farm animals, and

comfortable resting area; (3) Freedom from pain, injury, and disease—by prevention or rapid diagnosis and treatment; (4) Freedom to express normal behaviour—by providing sufficient space, proper facilities, and company of the animal's own kind; and (5) Freedom from fear and distress—by ensuring conditions which avoid mental suffering.

[7] TNA Webarchives, Farm Animal Welfare Council, Press Statement (05.12.1979), http://webarchive.nationalarchives.gov.uk/20121007104210/http://www.fawc.org.uk/pdf/fivefreedoms1979.pdf [19.12.2014], 2.

[8] For an overview of the wider political context see, Cassidy, *Vermin*, 48.

the legislation on experiments on live animals."[9] Meanwhile, Labour's manifesto promised bans of blood sports, a new council of animal welfare, and "stronger control on the export of live animals for slaughter, and conditions of factory farming, and experiments on living animals."[10]

Following Thatcher's 1979 election victory, winds of change were quickly felt within MAFF. Ministry officials were no longer in direct control of Britain's main welfare advisory body and complained that some members of a two-year enquiry into animal welfare by the House of Commons Agriculture Committee between 1980 and 1981 took "the view that the Brambell recommendations are Holy Writ."[11] The gradual weakening of industry-friendly bastions like MAFF was facilitated by a prolonged economic crisis of British agriculture. Starting in the late 1970s and gathering steam during the 1980s, many British livestock producers suffered from the joint effects of overproduction and stagnating or falling demand for animal products. Publicly, producers and agricultural officials also had to respond to damaging scandals involving drug residues, salmonellosis outbreaks, and the emerging mad cow disease (Bovine Spongiform Encephalopathy) crisis.[12] The combination of economic problems, scandals, and rising subsidies reduced support for MAFF positions within the free-market Thatcher administration.[13]

Corporatist bodies like MAFF were not the only ones to lose power over farm animal welfare politics. The events of 1979 also diminished the influence of former members of the RSPCA Reform Group. Already fiercely critical of FAWAC, the RSPCA Council had responded to the formation of the new FAWC with a boycott.[14]

[9] "1979 Conservative Party General Election Manifesto", http://www.conservativemanifesto.com/1979/1979-conservative-manifesto.shtml [02.02.2021]; for the earlier deleted version see: Margaret Thatcher Foundation, "Shadow Cabinet: Circulated Paper, The Conservative Manifesto 1978 – The Right Approach to Government. 2nd LCC Draft. Copy No. 6," 18–19, https://c59574e9047e61130f13-3f71d0fe2b653c4f00f32175760e96e7.ssl.cf1.rackcdn.com/8C1B6421465247B4BB8F6BE90097961B.pdf [01.04.2020].

[10] "1979 Labour Party Manifesto. The Labour Way is the Better Way", http://www.labour-party.org.uk/manifestos/1979/1979-labour-manifesto.shtml [01.09.2020].

[11] TNA MAF 369/272 Minute CH Shillito to Mr Steel (17.03.1981); House of Commons, *First Report from the Agriculture Committee, Animal Welfare in Poultry, Pig and Veal Calf Production, Session 1980–1981, 02.07.1981* (London: House of Commons, 1981).

[12] Kirchhelle, *Pyrrhic Progress*, 232–233; 240–246.

[13] Winter, *Rural Politics*, 138.

[14] Hugh Clayton, "Strains in the RSPCA worse after heated debate", *Times*, 25.02.1980, 16; Hugh Clayton, "Expulsion call threatens RSPCA board", *Times*, 19.11.1979, 14; Hugh Clayton, "Pressure mounts for RSPCA reforms", *Times*, 26.11.1979, 14.

What had triggered this decision? In early 1979, Thatcher's predecessor, Labour Prime Minister James Callaghan, had already proposed a new general oversight committee for all advisory committees on farm animal welfare, animal experimentation, and animal transportation. The new super committee, which was also mentioned in Labour's manifesto, would be tasked with recommending "changes in the law relating to animal welfare, in its administration and in the relevant advisory machinery."[15] Callaghan's decision had initially been welcomed by the RSPCA.[16] However, following Margaret Thatcher's victory at the May 1979 general election, the initially proposed super committee morphed into a much narrower FAWAC replacement. Ryder and his allies also became concerned by other activists' decision to turn down FAWC membership offers and the nomination of NFU-associated intensive farmers and Sydney Burgess, who was associated with firms exporting live animals.[17] In August 1979, the *Daily Star* claimed that the Society had warned the new Minister of Agriculture Peter Walker—a moderate Conservative with previous experience in setting up the Department of the Environment under the Heath government—that it would boycott FAWC if Burgess was appointed.[18] Facing parallel protests against MAFF badger-gassing,[19] Walker refused to give way.[20]

The RSPCA's resulting boycott of FAWC led to a further éclat when the Society's executive director Julian Hopkins and chief veterinary officer Peter Brown accepted FAWC appointments without consulting RSPCA

[15] British Library, Richard Ryder Papers, Ryder Dep. 9846, B3/1, Farm Animal Welfare Council, 1976–1980, Prime Minister Callahan to Ryder, 23.04.1979; "1979 Labour Party Manifesto. The Labour Way is the Better Way", http://www.labour-party.org.uk/manifestos/1979/1979-labour-manifesto.shtml [01.09.2020].

[16] British Library, Richard Ryder Papers, Ryder Dep. 9846, B3/1, Farm Animal Welfare Council, 1976–1980, Prime Minister Callahan to Ryder, 23.04.1979.

[17] British Library, Richard Ryder Papers, Ryder Dep. 9846, B3/1, Farm Animal Welfare Council, 1976–1980, Eileen Bezat to Ryder, 14.08.1979; Daily Star clipping, 08.08.1979; Bezat to Ryder, 17.11.1979; List of FAWC members [NFU members marked in red]; Garner, *Animals, Politics, and Morality*, 57–58; Ryder, *Animal Revolution*, 184–185.

[18] British Library, Richard Ryder Papers, Ryder Dep. 9846, B3/1, Daily Star Clipping, 08.08.1979. Cassidy, *Vermin*, 87, 97.

[19] Cassidy, *Vermin*, 87.

[20] British Library, Richard Ryder Papers, Ryder Dep. 9846, B3/1, Farm Animal Welfare Council, 1976–1980, *Daily Star* clipping, 08.08.1979; Copy of letter from Peter Walker to Janet Fookes, MP, 10.09.1979.

leadership. Both had to be ordered to lay down their mandates with a Council majority of one in November 1979.[21] Publicly justifying the RSPCA's boycott, Richard Ryder claimed:

> There were hardly any well known welfare campaigners on [FAWC]. You will not get progressive reforms from such a committee. ... It is a well known device in political circles to set up a committee to slow down progress. ... By supporting such a committee we reduce our opportunities to speak to the Government direct.[22]

The FAWC boycott exacerbated already significant rifts between Ryder's reform camp and a growing number of internal critics, who bemoaned the Society's alleged take-over by a radical minority. In addition to weakening Ryder's position and triggering damaging news coverage,[23] the boycott also enabled agricultural commentators to criticise the Society.[24] As the only prominent welfare representative left on FAWC, Harrison added to public pressure on Ryder. Writing to the *Times*, she criticised the RSPCA's impudence:

> The RSPCA council could learn much from FAWC in being able to differ in a friendly and civilized way. ... For the RSPCA to boycott [FAWC] would be incredibly foolish. You are not hurting FAWC, you are hurting the RSPCA and farm animals.[25]

Harrison's position was shared by a majority of RSPCA members. On February 23, 1980, an extraordinary general meeting debated the RSPCA

[21] British Library, Richard Ryder Papers, Ryder Dep. 9846, B3/1, Farm Animal Welfare Council, 1976–1980, RSPCA Press Release, 02.11.1979, Ryder, *Animal Revolution*, 184–85.

[22] Hugh Clayton, "Strains in the RSPCA worse after heated debate", *Times*, 25.02.1980, 16.

[23] Hugh Clayton, "Strains in the RSPCA worse after heated debate", *Times*, 25.02.1980, 16; Hugh Clayton, "Expulsion call threatens RSPCA board", *Times*, 19.11.1979, 14; Hugh Clayton, "Pressure mounts for RSPCA reforms", *Times*, 26.11.1979, 14.

[24] Peter Bell, "Animal Welfare Showdown Nears", *British Farmer* 12.01.1980, 20; "RSPCA spurns offer of land farm husbandry research", *British Farmer*, 02.02.1980, 25; "RSPCA prepares for night of the long knives", *British Farmer*, 16.02.1980, 14; "Animal welfare: NFU declares war", *British Farmer*, 01.03.1980, 25; "RSPCA Cauldron Still Simmering", *British Farmer*, 15.03.1980, 13; "Power Struggle Tears Cruel Rift in RSPCA", *British Farmer*, 05.07.1980, 18.

[25] Hugh Clayton, "Strains in the RSPCA worse after heated debate", *Times*, 25.02.1980, 16.

boycott but failed to generate the 60 per cent of votes required for expelling reform members.[26] The vote nonetheless marked a turning point for the Society's 1970s' reform movement. Although the 1980s witnessed further radical animal rights activism inside and outside the RSPCA,[27] the Society returned representatives to FAWC after a newly elected more moderate Council voted to end the boycott in July 1980.[28]

The failure of the RSPCA boycott was also a sign that FAWC was proving far more effective than its predecessor. Between 1979 and its replacement by a new Farm Animal Welfare Committee in 2011, FAWC published over 30 reports on animal welfare.[29] In contrast to FAWAC, FAWC's composition and greater independence from MAFF enabled it to conduct rapid welfare code reviews. Soon after its establishment, FAWC reviewed pig codes. Non-binding recommendations reflected "the Committee's belief that the keeping of sows and gilts in stalls, with or without tethers, gives rise to abnormal behaviour and very commonly causes injuries."[30] Further important reviews followed. Drawing on John Webster's FACT-sponsored strawyard experiments, a review of veal production led to legislation requiring digestible fibre in calves' diets and improved crate sizes—"effectively destroy[ing]"[31] remaining intensive production and leading to a 1986 ban of individually penned calf crates.[32] FAWC also tried to proactively improve farm animal welfare and published 117 recommendations for farm animals at the time of slaughter in 1984. In 1992, a FAWC review of egg production laid the ground for a gradual reduction of battery cage use and improvements of free range systems. In 1993, another review criticised harmful breeding practices and laid out welfare research aims.[33] Although most codes remained non-binding and actual

[26] British Library, Richard Ryder Papers, Ryder Dep. 9846, B3/1, Farm Animal Welfare Council, 1976–1980, Veterinary Record, 01.03.1980, "RSPCA – a deeply troubled body".

[27] Roscher, *Königreich*, 296–298, 419–496.

[28] Garner, *Animals, Politics, and Morality*, 57–58; "RPSCA patches over the cracks", *British Farmer*, 19.07.1980, 15.

[29] FAWC, *Farm Animal Welfare in Great Britain: Past, Present and Future* (London: FAWC, 2009), 10.

[30] TNA MAF 369/272 Background to Question 11, enclosed in: Handwritten list of questions and answers to enquiry by House of Commons Agriculture Committee, 2; the practice was eventually banned in 1999.

[31] Webster, *Cool Eye*, 188.

[32] FACT Files, DB, Unmarked Blue Ryman Folder, Intensive Farming Review (February 1987); FACT Files, DB, Fund Raising, Ruth Harrison to Mrs Miloe (31.08.1994).

[33] Webster, *Cool Eye*, 163–165, 181, 240.

legislation was often slow to materialise,[34] even former critics like Richard Ryder later acknowledged that FAWC had produced "a succession of sensible proposals for reform."[35]

In the meantime, welfare politics were becoming increasingly European. Although it often took a long time for European initiatives to transform into national regulations,[36] the Council of Europe had turned into an influential forum for welfare politics and inspired farm animal welfare engagement by the European Economic Community (later EU).[37] In addition to strengthening ties between European scientists, welfare organisations, and regulators, Council of Europe deliberations created significant peer pressure for member states to align standards. Resulting achievements were impressive. Starting in the 1970s, European committees laid the ground for the 1976 Council Convention for the Protection of Animals as well as for other European welfare regulations, including requiring the stunning of animals before slaughter (1974); incorporating FAWC's five freedoms into the Council's convention for the protection of animals kept for farming purposes (1978); approving the convention for the protection of animals for slaughter (1988); providing minimum standards for the welfare of laying hens in battery cages (1988, amended 1999 and 2002); providing marketing standards for eggs (1990); protecting animals during transport (1991, amended 1995, 1997, 2001); establishing minimum standards for the protection of calves (1991, amended 1997) and pigs (1991, amended 2001); and passing the 1997 Treaty of Amsterdam, which granted animals special legal consideration as sentient

[34] Welfare regulations after 1978 include the 1978 Welfare of Livestock (intensive Units) Regulations; the 1982 (Prohibited Operations) Regulations; 1983 welfare codes for cattle; 1987 (Welfare of Battery Hens) & (Welfare of Calves) Regulations; 1990 (Welfare of Livestock/ Welfare of Animals at Market) Regulations; 1991 (Welfare of Pigs) Regulations; and enhanced welfare for the keeping of battery hens, calves, pigs with the 1994 Animal Prevention of Cruelty. Welfare of Livestock Regulations. As a result of the BSE crisis, live British calf exports stopped between 1996 and 2006 and never fully recovered afterwards although the RSPCA raised concerns about animals having to travel longer distances in the UK due to declining slaughterhouse numbers; FAWC, *Farm Animal Welfare in Great Britain*; Food Ethics Council and Heather Pickett, *Farm Animal Welfare, Past, Present and Future* (Southwater: RSPCA, 2014), 8–9; Webster, *Cool Eye*, 260.

[35] Ryder, *Animal Revolution*, 185.

[36] Millman et al., "The impact of applied ethologists," 300.

[37] Broom, "World Impact of ISAE," 271–273.

beings.[38] The EU also passed a 2001 convention banning sow and tether stalls by 2013, following a unilateral ban in Britain in 1999.[39]

For Ruth Harrison, the 1980s marked the highpoint of her career as a full-time campaigner. Within FAWC, she no longer had to defend basic principles of positive welfare, uphold behavioural perspectives, or block industry-friendly standards. Instead, she could concentrate on proactively shaping code improvements. At the European level, she exerted influence as the World Society for the Protection of Animals (WSPA) representative within the Standing Committee of the European Convention for the Protection of Animals Kept for Farming Purposes (T-AP).[40] She also became a prominent member of the Eurogroup for Animal Welfare, which had been founded as a European welfare lobby organisation in 1980.[41] Her success in maintaining access to both European and British decision-making bodies did not mean that she became less combative. According to fellow FAWC and T-AP members, meetings could still be characterised by Ruth Harrison acting as a "minority of one."[42]

The 1980s were also a time of heightened FACT activity, matching Harrison's involvement in FAWC, T-AP, and the Eurogroup. The Trust continued to focus on providing supportive research for Harrison's FAWC work. During the early 1980s, FACT and the UFAW jointly financed a further Gallup Poll to find out whether consumers would pay more for non-battery eggs: "The survey had indicated that nearly two-thirds of consumers would be prepared to pay over 5p/dozen more for non-battery eggs."[43] Speaking to trustees in September 1983, Ruth Harrison gave an overview of FACT projects: The Bristol veal project had been completed, a FACT-supported report on animals and ethics was selling well,[44] and FACT had agreed to support research by David Wood-Gush and Alex

[38] Millman et al., "The impact of applied ethologists," 301.

[39] Food Ethics Council and Heather Pickett, *Farm Animal Welfare*, 9.

[40] Broom, "Ruth Harrison's Later Writings," 22.

[41] Eadie, *Understanding Animal Welfare*, 24.

[42] Oral History Interview Marian Stamp Dawkins (01.07.2014).

[43] FACT Files, MD, Minute Book, FACT, Minutes of Meeting of Trustees (14.05.1981—signed by Ruth Harrison on 27.09.1983), 1.

[44] The title of the report was *Animals and Ethics*, FACT Files, DB, FACT 93/94, FACT. Report for the Year Ended June 30, 1994, 1; FACT Files, MD, Minute Book, FACT, Minutes of Meeting of Trustees (27.09.1983); the book had also been introduced to the Yearly Meeting of the Society of Friends by former Brambell member and FACT trustee William Thorpe; FACT Files, MD, Minute Book, FACT, Minutes of Meeting of Trustees (14.05.1981—signed 27.09.1983), 2.

Stolba on alternative 'family pen' systems for pigs at the University of Edinburgh.[45] The Trust was also financing a film on pig behaviour, sponsoring a Behavioural Needs workshop, and was considering establishing a journal on farm animal care.[46] In the same year, FACT also decided to support research on low-cost free range systems for laying hens[47] and approached corporations to promote humane slaughtering and stunning devices.[48]

Fundraising and publicity activities increased in tandem with FACT's research sponsorship. At the 1982 and 1984 Royal Shows, FACT collaborated with the Royal Agricultural Society of England and mounted an exhibit titled *Farm Animals: Towards Alternative Systems.* In 1984, it established the so-called FACT Award "for the design and production of a machine for the humane collection and transport of broiler chickens for slaughter."[49] Although it failed to secure sponsorship by the Prince of Wales and Paul and Linda McCartney,[50] FACT managed to attract the patronage of the renowned natural scientist Dame Miriam Rothschild.[51] Following the resignation of FACT trustee and former FAWAC and FAWC chair Prof Richard Harrison,[52] Ruth Harrison convinced prominent younger welfare researchers, including Donald Broom and—despite earlier differences—Marian Dawkins, to conduct FACT-supported research and become trustees.[53]

All the while, FACT's agenda remained almost single-handedly determined by its chairwoman, who used trust funds to support her work on

[45] FACT Files, MD, Minute Book, FACT, Minutes of Meeting of Trustees (27.09.1983), 1.

[46] FACT Files, MD, Minute Book, FACT, Minutes of Meeting of Trustees (27.09.1983), 1–2.

[47] FACT Files, DB, Trustees Meetings, Farm Animal Care Trust [undated], point three.

[48] FACT Files, MD, FACT, Minutes of a Meeting of Trustees (27.09.1983), 1.

[49] FACT Files, MD, FACT Publications & Publicity Material, Pamphlet FACT, 4.

[50] FACT Files, DB, Unmarked Green Ryman Folder, Harold Rose to potential sponsors (22.04.1984), Harold Rose to HRH The Prince of Wales (08.10.1984), enclosed in: Harold Rose to Ruth Harrison (17.05.1987); FACT Files, DB, Unmarked Blue Ryman Folder, Su Gold to Ruth Harrison (27.03.1984); FACT Files, DB, Unmarked Green Ryman Folder, Ruth Harrison to Richard (17.06.1985).

[51] FACT Files, DB, Unmarked Green Ryman Folder, Miriam Rothschild to Ruth Harrison (14.07.1984).

[52] FACT Files, DB, Trustees Meetings, Richard Harrison to Ruth Harrison (01.09.1985).

[53] FACT Files, DB, Unmarked Red Ryman Folder, Don Broom to Ruth Harrison (20.03.1986); FACT files, DB, Unmarked Green Ryman Folder Marian Dawkins to Ruth Harrison (25.06.1985).

British and European animal welfare committees and travel to scientific meetings.[54] During the second half of the 1980s, FACT also sponsored further research on improved stunning and slaughtering devices for cattle and poultry, alternative production systems for animals, and the space needs of laying hens.[55] As usual, research sponsorship was closely tied to political campaigning. In a letter to Miriam Rothschild from 1985, Harrison listed her hopes for the immediate future of British animal welfare:

> If there were regulations giving all animals the 'five freedoms' and a well bedded lying area, this would eliminate at one stroke veal crates, sow stalls, tie stalls, piglet cages and flat deck cages, battery cages for laying hens, rabbit cages and all the extreme systems. Add a diet to keep the animal in full health and vigour and 'quality veal' is out as well.[56]

Commenting on recent FAWC work on animal transports, Harrison did not think that quick regulatory progress was likely but wryly noted that there was "no harm in [external] pressure," such as research and publicity generated by FACT, as it would make "FAWC's job easier" (Image 11.1).[57]

Other reforms advocated by Harrison and supported by FACT included a welfare-based rating and licensing system for farm buildings and stockmen, a mandatory stock to stockman ratio, the provision of well-bedded lying areas for stock and of perches and nesting boxes for birds, requirements for feeds to preserve animals' full health and vigour, and ad-lib animal access to water.[58] FACT also funded research on poultry, turkey, and pig housing; the force-feeding of ducks and geese for *foie gras*; and improved pre-slaughter stunning. In the latter case, Harrison tested gas stunning via CO_2 (with 2 per cent oxygen) and electro-immobilization on

[54] FACT Files, DB, Unmarked Green Ryman Folder, David Wood-Gush to Ruth Harrison [undated].

[55] FACT Files, MD, Minute Book, FACT, Minutes of a Meeting of Trustees (10.10.1986); The Farm Animal Care Trust. Current & Completed Projects [undated].

[56] FACT Files, DB, Unmarked Green Ryman Folder, Ruth Harrison to Dr Rothschild (03.06.1985).

[57] FACT Files, DB, Unmarked Green Ryman Folder, Ruth Harrison to Dr Rothschild (03.06.1985).

[58] FACT Files, DB, Unmarked Green Ryman Folder, Farm animals—some suggested improvements, enclosed in: Ruth Harrison to Dr Rothschild (03.06.1985).

Image 11.1 Ruth Harrison and Klaus Vestergaard observe a sow at the Swedish Pig Park in 1988 (image courtesy of Bo Algers)

herself. She was not convinced by either technology.[59] With public protests against live animal transports flaring up again,[60] Harrison advocated slaughtering animals close to their place of production, a ban on live animal exports, and improvements in domestic slaughtering arrangements.[61] In 1985, she accused the government of delaying the implementation of

[59] Van De Weerd and Sandilands, "Bringing the Issue of Animal Welfare to the Public," 408; Webster, "Ruth Harrison," 8.

[60] Howkins and Merricks, "'Dewy-Eyed Veal Calves'."

[61] FACT Files, DB, Unmarked Green Ryman Folder, Farm animals—some suggested improvements, enclosed in: Ruth Harrison to Dr Rothschild (03.06.1985).

new 1984 FAWC recommendations for humane slaughtering.[62] Three years later, she renewed her long-standing criticism of legal pre-stunning exemptions for so-called religious ritual slaughter.[63]

Despite experiencing a personal blow through the death of her husband, Dex, in December 1987,[64] Ruth Harrison intensified the time and effort she devoted to farm animal welfare. A sudden influx of money helped her do so. Although fundraising had improved since the 1970s, FACT's net assets had not exceeded £10,000–20,000. However, in 1988, donations of £40,428 significantly improved the Trust's budget.[65] In addition to other forms of income such as reports on animal welfare on individual farms (see Chap. 12),[66] re-investing the donations enabled FACT to professionalise and expand sponsorship of ethological research.[67] Having already funded a conference on the behavioural needs of farm animals in 1987, FACT co-financed additional meetings on the economic viability of humane production systems in 1991 and on sustainable livestock production in 1993.[68] FACT also began contributing to scientists' travel expenses and launched new research projects such as a "high welfare pig building project"[69] at the Edinburgh School of Agriculture. In 1988, FACT established a scholarship on the "feasibility of combining pigs/poultry with trees"[70] in memory of Dex Harrison. During the early 1990s, FACT co-sponsored a design competition for cattle stunning pens and financed research on pig farrowing systems, different methods of stunning and euthanising poultry, sheep housing in winter, fish slaughter, animals' space needs, alternative husbandry systems, and the microchip

[62] FACT Files, DB, Unmarked Green Ryman Folder, Farm animals—some suggested improvements, enclosed in: Ruth Harrison to Dr Rothschild (03.06.1985).

[63] N.C. Sweeney, "Animal welfare and ritual slaughter", *Times*, 14.06.2003, 27.

[64] "Dex Harrison – Basic Biographical Details".

[65] FACT Files, DB, FACT 89/90, Draft: Farm Animal Care Trust, Year Ended 30.06.1989; FACT Files, DB, FACT 90/91, Draft Proposals for consideration by the Farm Animal Care Trust (04.02.1989). It is unclear who made the large donations.

[66] FACT Files, DB, Fund Raising, Sheet—FACT Income July 1989 to end of July 1991.

[67] FACT Files, DB, FACT 89/90 P.R. Lansberry to L.F. Hawken (10.04.1989).

[68] FACT Files, MD, FACT Publications & Publicity Material, Pamphlet [undated, probably post-2000], 2.

[69] FACT Files, DB, Unmarked Green Ryman Folder, Colin T. Whitemore to Ruth Harrison (13.06.1988); FACT Files, DB, Unmarked Red Ryman Folder, Bryan Jones to Ruth Harrison (15.11.1988).

[70] FACT Files, MD, FACT Publications & Publicity Material, The Farm Animal Care Trust [pamphlet], 2–3.

feeding of pigs.[71] Further plans centred on publicly promoting Trust-sponsored research.[72]

However, despite her nomination to FAWC, the expansion of FACT, and her 1986 elevation to the status of Officer of the Most Excellent Order of the British Empire (OBE),[73] Ruth Harrison remained dissatisfied with the overall state of British welfare. In 1988, she complained that widespread societal acceptance of "Professor Thorpe's 1965 guiding principle[s]"[74] had not led to a ban of problematic production systems. The prefaces to the UK's 1971 Codes of Practice had "studiously ignored behaviour, and related welfare only to physiological requirements."[75] Despite signing the 1976 Council of Europe Convention, the UK had passed new welfare regulations only relating to the daily inspection of housed livestock and automated equipment—"everything else, [officials] felt was covered by the 1968 Agriculture (Miscellaneous Provisions) Act."[76] Scandinavian countries and West Germany had outpaced the UK in terms of welfare regulation. To maintain the wider status quo of welfare regulations, British officials were clinging to the European Convention's

[71] FACT Files, MD, FACT Publications & Publicity Material, The Farm Animal Care Trust [pamphlet]; FACT Files, DB, Unmarked Green Ryman Folder, Clive Hollands to Ruth Harrison (21.04.1990); FACT—Current Projects; FACT Files, DB, Unmarked Red Ryman Folder, A device to monitor the operation of electrical stunners—a report commissioned by FACT (19.06.1990).

[72] FACT Files, DB, Unmarked Green Ryman Folder, FACT—Current Projects, 3; FACT Files, DB, Unmarked Red Ryman Folder, FACT and Agricultural and Food Research Council. Institute of Food Research. Innovation Agreement (Sept. 1991), 2.

[73] Ryder, "Harrison, Ruth (1920–2000)".

[74] FACT Files, MD, FACT Publications & Publicity Material, Ruth Harrison, "Introduction – Proceedings of Workshop sponsored by the Farm Animal Care Trust and the Universities Federation for Animal Welfare: Behavioural needs of Farm Animals," *Applied Animal Behaviour Science* 19 (1988), 341.

[75] FACT Files, MD, FACT Publications & Publicity Material, Ruth Harrison, "Introduction – Proceedings of Workshop sponsored by the Farm Animal Care Trust and the Universities Federation for Animal Welfare: Behavioural needs of Farm Animals," *Applied Animal Behaviour Science* 19 (1988), 341.

[76] FACT Files, MD, FACT Publications & Publicity Material, Ruth Harrison, "Introduction – Proceedings of Workshop sponsored by the Farm Animal Care Trust and the Universities Federation for Animal Welfare: Behavioural needs of Farm Animals," *Applied Animal Behaviour Science* 19 (1988), 343.

call for the qualification of welfare regulations through "established experience and scientific knowledge" "like drowning men to a straw."[77]

Much remained to be done. The next decade, however, saw Ruth Harrison's campaigning career slowly come to an end. Celebrating her 70th birthday in 1990 and reaching her maximum term of office in 1991, Harrison left FAWC. Her retirement marked the end of 24 years of membership on Britain's leading welfare committees.[78] Despite her retirement, Harrison maintained a degree of influence on British and European animal welfare debates. In 1990, a *Guardian* article described Ruth Harrison and Rachel Carson as two "solitary prophets"[79] of the twentieth-century animal welfare and environmentalist movements. Glossing over Harrison's actions during the turbulent 1970s and drawing heavily on clichés of 'sentimental' female activism, the *Guardian* claimed that Harrison's campaign for animal welfare had been characterised by her "moderate views and step-by-step approach."[80] Speaking with "a gentle voice"[81] to the article's author, Harrison expressed understanding for post-war governments' attempts to boost meat production. However, she remained adamant that the production methods chosen had been wrong. When it came to the intensive production of white veal, Harrison grew agitated: "For the first time she raises her voice. 'Why white veal? Why the hell white veal?'"[82] Summarising the evolution of welfare regulations since *Animal Machines* and contrasting her with more radical campaigners, the article expressed certainty that Harrison's "Tolstoy-like" strategy of incremental improvements would continue to improve farm animals' welfare.[83]

This prediction proved true. Aided by FACT funding and her status as an iconic campaigner, Harrison continued to push for improvements in all areas of farm animal welfare.[84] It was only after a cancer diagnosis in 1996

[77] FACT Files, MD, FACT Publications & Publicity Material, Ruth Harrison, "Introduction – Proceedings of Workshop sponsored by the Farm Animal Care Trust and the Universities Federation for Animal Welfare: Behavioural needs of Farm Animals," *Applied Animal Behaviour Science* 19 (1988), 342.

[78] FACT Files, DB, Fund Raising, Ruth Harrison to Mrs Miloe (31.08.1994); Oral History Interview Donald Broom (04.07.2014).

[79] Colin Spencer and Spike Gerrel, "A rare breed at the factory farm", *Guardian*, 03.11.1990, A19.

[80] Spencer and Gerrel, "A rare breed at the factory farm".

[81] Spencer and Gerrel, "A rare breed at the factory farm".

[82] Spencer and Gerrel, "A rare breed at the factory farm".

[83] Spencer and Gerrel, "A rare breed at the factory farm".

[84] FACT Files, DB, FACT 93/94, Ruth Harrison, Farm Animal Care Trust, Report of the Trustees, Year Ended 30.06.1993, 1.

that FACT activities began to decline.[85] Still regularly attending T-AP's Strasburg meetings and re-arranging her chemotherapy so as not to clash with trips,[86] Harrison focused her final campaigning on securing new welfare guidelines for ducks, ratites, and pheasants. She also remained concerned about the use of carbon dioxide to cull animals like mink and commissioned research on alternative slaughter methods and lighting levels on farms (Image 11.2).[87]

By the end of the millennium, few of these demands were considered radical. Thirty-six years after the publication of *Animal Machines*, nearly all of the fundamental animal welfare positions espoused by Harrison had become part of mainstream culture and politics. In 1997, all major British parties included animal welfare statements in their election manifestoes. Pointing to the 1995 EU regulations on animal transports and the 1996 EU ban of veal crates in the middle of the contemporary mad cow disease crisis (see Chap. 12), the Conservative Party promised to "continue to take the lead in improving standards of animal welfare in Europe."[88] Meanwhile, Labour pledged to hold a free vote on whether to ban hunting with hounds.[89] The mainstreaming of her formerly radical positions was a sign of Harrison's success as a bestselling author, as a determined force within welfare committees, and as a veteran campaigner with a well-developed network of ties to leading scientists and decision-makers. It was also a sign that misogynist 1960s attempts to downplay her positions as overly emotional and later descriptions of Harrison as too timid to effect change had clearly been misplaced. By focusing on all stages of

[85] Oral History Interview Donald Broom (04.07.2014); FACT Files, DB, FACT 95/96, Farm Animal Care Trust, Year Ended 30.06.1995; Farm Animal Care Trust, Year Ended 30.06.1996; FACT Files, MD, Farm Animal Care Trust, Report for the Year Ended 20.06.1997; FACT Files, MD, Annual Returns 1998, Farm Animal Care Trust, Year Ended 30.06.1998.

[86] Oral History Interview Donald Broom (04.07.2014); Oral History Interview Ruth Layton (02.07.2014).

[87] FACT Files, MD, FACT 96/97, Report 1997; FACT Files, DB, FACT 93/94, FACT. Report for the Year Ended June 30, 1994, 1; FACT Files, MD, Annual Returns 1998, Farm Animal Care Trust, Year Ended 30.06.1998.

[88] *You can only be sure with the Conservatives*, 1997 Conservative Party General Election Manifesto, http://www.conservative-party.net/manifestos/1997/1997-conservative-manifesto.shtml [15.04.2020].

[89] *New Labour Because Britain Deserves Better*, 1997 Labour Party General Election Manifesto, http://www.labour-party.org.uk/manifestos/1997/1997-labour-manifesto.shtml [15.04.2020].

Image 11.2 Ruth Harrison at a Danish mink farm in 1997 (image Courtesy of Marlene Halverson)

Harrison's career rather than individual moments like the 1964 publication of *Animal Machines* or 1970s clashes with Reform Group members, we come to appreciate her remarkable ability to successfully negotiate a wide range of evolving political and campaigning environments. Many other activists, scientists, and politicians had shared overlapping ethical and scientific beliefs, but very few had been able to consistently influence British and European developments for over three decades.

CHAPTER 12

Non-conform Evidence: The Impasse of 1990s Welfare Research

The political mainstreaming of welfare issues was a boon for animal welfare science. Founded during the 1960s and navigating a tumultuous political marketplace during the 1970s, animal welfare science rapidly institutionalised during the 1980s and 1990s: funding levels and publications increased, welfare researchers obtained chairs at prestigious universities, and the discipline gained greater political influence on British and European decision-making bodies. Mirroring the rise of organic agriculture,[1] farm assurance schemes and quality-assured welfare labels presented a second important way of influencing welfare standards.[2] However, despite scientists' improved influence and resources, fundamental questions about how to define and measure welfare remained open. Previously favoured welfare definitions were challenged by non-conforming results including stereotyped 'abnormal' behaviour in healthy animals, stress in animals voluntarily performing 'natural' behaviour like mating, or sub-clinical disease in 'normally' behaving animals.[3] While most researchers remained confident in their ability to produce meaningful results,[4] animal welfare science entered a prolonged phase of epistemic navel-gazing.

[1] Kirchhelle, *Pyrrhic Progress*, 244–245.

[2] Food Ethics Council and Heather Pickett, *Farm Animal Welfare*, 10.

[3] Ian J. H. Duncan, "Science-based assessment of animal welfare: farm animals," *Revue scentifique et technique – Office International Epizooties* 24 (2) (2005), 483–484.

[4] Millman et al., "The impact of applied ethologists," 306–308.

C. Kirchhelle, *Bearing Witness*, Palgrave Studies in the History of Social Movements,
https://doi.org/10.1007/978-3-030-62792-8_12

In 2008, senior researcher David Fraser noted that its value-laden character made it doubtful whether welfare could ever be formally defined.[5]

Fraser's statement is indicative of the epistemic challenges and increasing heterodoxies faced by most expanding disciplines like that of classic ethology around 1970 (Chap. 10). Although *Web of Science* is far from exhaustive, does not include official reports, and underrepresents pre-digital and non-English contributions, a search for publications mentioning 'animal welfare' indicates a surge of outputs from the 1980s onwards. The 1970s had seen an average of seven dedicated welfare publications per year.[6] This number increased over fourfold to 33.5 publications per year during the 1980s. By the 1990s, it had risen by a further fivefold to an average of 183.9 publications per year. Ahead of the US (3513, 17.7 per cent of 19,838 publications) and Germany (2048, 10.3 per cent of 19,838 publications), Britain (England, Wales, Scotland, Northern Ireland) contributed the most publications (3835, 19.3 per cent of 19,838 publications) with the Universities of Bristol (538), Edinburgh (337), London (317), Oxford (204), Newcastle (202), and Cambridge (200) emerging as the most significant research hubs.[7]

The surge of publications was paralleled by a further institutionalisation of the veterinary and behavioural welfare sciences. Founded in 1966 and opened to non-veterinarians from 1970 onwards, the Society for Veterinary Ethology began hosting international meetings from the 1970s onwards and was renamed International Society for Applied Ethology (ISAE) in 1991. By 2003, the ISAE had 729 members from around the world. While the ISAE's *International Journal of Applied Behavioural Sciences* (formerly ethology) remained influential, new journals like the UFAW's *Animal Welfare* (est. 1992) were created to serve the growing needs of the community.[8] Animal welfare science's expanding influence was reflected in Britain's university landscape. In 1977, John Webster was appointed to Bristol's Chair of Animal Husbandry. In the same year, Marian Dawkins obtained a permanent lectureship (by 1998, a professorship) in zoology at the University of Oxford. In 1986, the University of

[5] Fraser, "Understanding Animal Welfare," *Acta Veterinaria Scandinavica* 50/ Supplement (2008), S1.

[6] An exceptional 28 publications appeared in 1979.

[7] Web of Science, "Web of Knowledge" search of term "animal welfare" [05.04.2020].

[8] Millman et al. "The impact of applied ethologists," 300 & 309; Petherick and Duncan, "The International Society for Applied Ethology," 34, 39–47.

Cambridge's Department of Clinical Veterinary Medicine appointed Donald Broom to the first dedicated chair of animal welfare science.[9]

It would, however, be wrong to think of animal welfare science as a phenomenon limited to universities and political committees. As described by Emma Roe and Henry Buller, the field's growth was also aided by the rise of welfare as an economic value.[10] Although radical protesters continued to oppose intensive farming per se,[11] the mainstreaming of welfare values (Chap. 11) led to a mutually beneficial cooperation between researchers, industry, and established animal charities.

By the late 1980s, major retailers and producers began to see farm animal welfare not only as a factor whose absence might impede productivity but also as a value whose certifiable presence might boost sales on the relatively homogeneous market for animal products.[12] The demand-led trend towards the value-based segmentation of the British food market was already evident in the consistent growth of sales of 'naturally' or 'organically' produced food, which was now on offer in major supermarkets like Safeway, Waitrose, Sainsbury's, Tesco, and Marks & Spencer.[13] It was also affecting environmental and welfare practices on conventional farms. In the poultry sector, consumer preferences drove a gradual shift of egg production from battery cages to free range and deep litter systems. In pig production, consumer demand and lower costs made a substantial number of producers adopt straw-based indoor or extensive outdoor systems.[14]

Because it was impossible to 'see' added ethical or health values in a product, a thriving certification industry emerged to aid the growth of

[9] "UFAW Medal for Outstanding Contributions to Animal Welfare Science – past awards", *Universities Federation for Animal Welfare*, https://www.ufaw.org.uk/ufaw-medal-for-outstanding-contributions-to-animal-welfare-science/ufaw-medal-for-outstanding-contributions-to-animal-welfare-science-past-awards [13.04.2020].

[10] Buller and Roe, *Food and Animal Welfare*, 49–51.

[11] Tester, "The British Experience," 241–251; Howkins and Merricks, "'Dewy-Eyed Veal Calves'," 85–103; Buller and Roe, *Food and Animal Welfare*, 45–46; Linda Merricks, "Green Politics," 437–442; Roscher, *Königreich*, 419–504.

[12] Buller and Roe, *Food and Animal Welfare*, 49–51; see also: Henry Buller, "Animal welfare: from production to consumption," in H. Blokhouis et al. (eds.), *Welfare quality: science and society improving animal welfare* (Wageningen: Wageningen Academic Press, 2013), 49–69.

[13] Kirchhelle, *Pyrrhic Progress*, 244.

[14] Food Ethics Council and Heather Pickett, *Farm Animal Welfare*, 8–10; Martin, *Development of Modern Agriculture*, 124.

premium segments of the food market. Certification and farm assurance schemes created a lucrative win-win-win alliance between three distinct actor groups: consumers who wanted to acquire and support the production of food with 'superior' ethical and health qualities; retailers and producers who wanted a means to designate and add financial value to food produced according to higher voluntary standards; and animal protection organisations that wanted to raise welfare standards and their own income by endorsing and policing specific practices via labels.[15] Exacerbating the 1980s' breakdown of agricultural corporatism (Chap. 11), the "virtuous bicycle"[16] of assurance schemes further shifted power away from official, producer-focused entities like MAFF towards privatised, consumer-oriented solutions in a marketplace that was increasingly dominated by large supermarkets and vertically integrated agribusiness.

The organic sector led the way. In 1973, the British Soil Association had already begun to certify that members were producing organic food according to strictly defined methods. This informal certification scheme was officially recognised by the 1987 UK Register for Organic Food Standards and the 1989 Organic Standards. European standardisation followed with the 1991 EEC Council Regulation (2092/91) on organic production of agricultural products.[17] Conventional producers also recognised the advantages of assurance and labelling schemes. Reacting to food scares and foreign competition, the British and Scottish governments attempted to boost sales of domestic products via assurance schemes like the Food from Britain scheme (1984–1993), the Scottish Livestock Assurance Schemes (1987), and the Quality Meat Scotland scheme (1991).[18]

Large-scale assurance schemes that specifically targeted animal welfare emerged in the 1990s. Inspired by the success of organic labelling and Audrey Eyton's *Kind Food Guide* (1991),[19] the RSPCA collaborated with retailers and producers to create its Freedom Foods Label in 1994. The move was backed by surveys, which found that 95 per cent of consumers favoured welfare labelling. MPs from all parties praised the fact that consumers would "for the first time (…) be offered a clear choice of meat and

[15] Buller, "Animal Welfare: from production to consumption"; Food Ethics Council and Heather Pickett, *Farm Animal Welfare*, 18–19.

[16] Food Ethics Council and Heather Pickett, *Farm Animal Welfare*, 19.

[17] Kirchhelle, *Pyrrhic Progress*, 244; Food Ethics Council and Heather Pickett, *Farm Animal Welfare*, 11.

[18] Food Ethics Council and Heather Pickett, *Farm Animal Welfare*, 11.

[19] Audrey Eyton, *The Kind Food Guide. Kinder to animals – much kinder to you* (London: Penguin, 1991).

dairy products that have been produced with high standards."[20] Non-statutory welfare labelling presented a "commercial opportunity for farmers who place a high regard for the care and protection of their animals" and would provide "a basis for a gradual and steady improvement in the welfare of Britain's 750 million farm animals."[21]

Freedom Foods was also an opportunity to deescalate confrontations on the public 'frontstage' of farm animal welfare politics. By defining the "consumer as a positive figure,"[22] politicians, industry, and the RSPCA opened a mutually beneficial 'backstage' debate about how welfare and market demands might meet. Access to this corporate backstage was restricted, and there was limited opportunity for protests. Consultations over what constituted welfare and how it could be assured were conducted by senior retailers, producers, RSPCA officials, and experts. Internal and external opposition against RSPCA engagement with industry could be deflected by referencing income being generated for RSPCA work in other areas, the fact that consumers wanted labels and that public campaigns for statutory change remained possible. Meanwhile, retailers offering Freedom Food could defend sales of foodstuffs produced with less stringent criteria by arguing that consumers were free to choose more expensive ethical products. According to Matthew Hilton and others, this notion of market-based citizenship "was consistent with the broader processes of the privatization of politics upon which NGOs sought to capitalize":

> Shopping was an opportunity for the NGO supporter to demonstrate commitment to the cause in a manner which also expressed loyalty to the message put forward by the NGO leadership. (…). What an engagement with consumption enabled was a disciplining of supporter behaviour for even the most passive sympathizer.[23]

Not everyone was happy with the shift of welfare politics and standard-setting towards the marketplace. Despite her previous calls for a consumer revolt, Ruth Harrison was unimpressed by the increasing emphasis on

[20] House of Commons, Early Day Motion No. 1543, "Freedom Food", Tabled 19.07.1994, 1993–1994 Session, https://edm.parliament.uk/early-day-motion/8714/freedom-food [01.05.2020].

[21] House of Commons, Early Day Motion No. 1543, "Freedom Food", Tabled 19.07.1994, 1993–1994 Session, https://edm.parliament.uk/early-day-motion/8714/freedom-food [01.05.2020].

[22] Hilton et al., *Politics of Expertise*, 215; on front- and backstage welfare politics see, *Cassidy, Vermin*, 205.

[23] Hilton et al, *Politics of Expertise*, 215–216.

individual choice and too-close alliances with industry. In 1994, she joined Joyce D'Silva (Compassion in World Farming) and Joanne Bower (Farm and Food Society) in criticising the RSPCA's 'Freedom Food' label in the *Times*. The Freedom Food label supposedly guaranteed farm animals' basic freedoms from fear, distress, pain, injury, disease, hunger, thirst, and discomfort. However, the three critics attacked its toleration of practices such as tail-docking, beak trimming, and sow stalls.[24] According to the *Observer*, Harrison was "frustrated because she feels something positive could have been achieved, but instead the RSPCA will betray the trust of consumers who place their confidence in the charity's name":

> It makes me profoundly unhappy, because I don't think they merit that trust. Their idea is you start with weak standards and improve them every so often. But once farmers have invested in a system, they're not going to change it every year. It's pie in the sky.[25]

Despite Harrison's criticism, assurance schemes with welfare elements continued to surge. In 1996, the mad cow disease (Bovine Spongiform Encephalopathy) crisis devastated Britain's beef industry and created a widespread moral panic about the health and ethical hazards of intensification.[26] Trying to restore trust, the government and industry designed new quality assurance schemes like the British Lion Scheme for eggs with compulsory vaccination against salmonella (1998) and the NFU's Red Tractor Scheme (2000).[27] Both initiatives proved popular among producers. Although critics periodically bemoaned weak inspection and welfare standards,[28] the Red Tractor Scheme covered nearly 100 per cent of UK-farmed salmon, 90 per cent of pigs and poultry, over 80 per cent of cattle, and 65 per cent of sheep in 2014. The premium welfare sector also expanded. Between 1994 and 2014, the number of British terrestrial farm

[24] D'Silva, Joyce et al., "Freedom food that fails the animals", *Times*, 09.07.1994, 19.

[25] "Consuming Passions", *Observer*, 10.07.1994, D10–11.

[26] Kirchhelle, *Pyrrhic Progress*, 225–226; on moral panics see Nicolas Rasmussen, "Goofball Panic: Barbiturates, 'Dangerous' and Addictive Drugs, and the Regulation of Medicine in Postwar America," in Jeremy A. Greene and Elizabeth Siegel Watkins (eds.), *Writing, Filing, Using, and Abusing the Prescription in Modern America* (Baltimore: Johns Hopkins University Press, 2012), 25.

[27] Food Ethics Council and Heather Pickett, *Farm Animal Welfare*, 11.

[28] See, for example, Martin Hickman, "The 'good food' stamp barely worth the label it's printed on", *Independent*, 01.05.2012; Ben Webster, "Red Tractor accepts need for change as shoppers want more spot checks", *Times*, 30.07.2018; "Flat House Farm pigs filmed living in 'barbaric conditions'", *BBC News* (24.08.2020).

animals raised according to RSPCA Freedom Foods standards rose from less than 100,000 to over 40,000,000. Major supermarkets also reacted to consumer demand by ending sales of cage-produced shell eggs.[29]

Growing demand for welfare assurance cemented the societal standing of animal welfare science and unlocked financial resources for researchers in academia and private certification bodies. However, it did not resolve ongoing disagreement about what welfare was. During the 1980s, increasingly sophisticated research on animal preferences, adaptive behaviour, and farm animals' physiology and neurobiology had moved welfare science well beyond early hormonal theories of stress, instinct concepts, and ideals of harmony with nature (Chap. 10).[30] However, researchers still found it hard to agree on universal welfare parameters, with some arguing that welfare was about how an animal felt and what it wanted and others arguing for predominantly adaptive physiological definitions of welfare.[31]

Disagreements about how to interpret different indicators affected the discipline's ability to establish coherent international welfare standards. While a 1997 European review of gestation stalls for sows resulted in a 2013 EU ban, an expert report on the same practice in Australia could not identify significant welfare problems. The contradictory outcome was in turn used by the US swine industry to argue that there was inconclusive evidence for the elimination of the stalls.[32] During legislative hearings in the US, veterinary and behavioural researchers also disagreed on whether laying hens suffered as a result of forced moulting.[33]

Disagreements about methods, indicators, and standards resulted in sustained debates about animal welfare science's methods and effectiveness. In 1995, John Webster's *Animal Welfare: A Cool Eye Towards Eden* launched a scathing attack on the limited scope of welfare regulations, the lack of evidence underpinning much of animal philosophy and activism,

[29] Food Ethics Council and Heather Pickett, *Farm Animal Welfare*, 3, 12, 14.

[30] Ian J.H. Duncan, "D.G.M. Wood-Gush Memorial Lecture: An applied ethologist looks at the question "Why?"," *Applied Animal Behaviour Science* 44 (1995), 205–217; Webster, *Cool Eye*; Marian Stamp Dawkins, "Why has there not been more progress in animal welfare research? – D.G.M. Wood-Gush Memorial Lecture," *Applied Animal Behaviour Sciences* 53 (1997), 59–73; Broom, "A History of Animal Welfare Science," 121–137.

[31] Broom, "A History of Animal Welfare Science," 127; Duncan, "Science-based assessment of animal welfare," 484–486.

[32] Fraser, "Understanding Animal Welfare," S1.

[33] Millman et al., "The impact of applied ethologists," 305.

and scientists' limited influence "on the quality of life for the vast majority of animals reared for food."[34] Although the animal welfare movement had had a significant impact on social values, single-sentence definitions of welfare remained inadequate and "a lot of very well-intended welfare research is neither very good science nor very helpful to the animals."[35] Webster was particularly sceptical of crude physiological and neurological welfare measures: "My particular bête noire is the experiment which seeks only to obtain a so-called 'objective' measure of something which the researcher preconceives to be stress."[36] For Webster, welfare definitions and politics should be based on a cost-benefit calculation of the "things we do to animals for our benefit," the "cost to us of acting for their benefit," and the cost to us "of breaking our current association with an animal species."[37] Scientists could not answer these issues by themselves but would have to take into account wider considerations of morality, politics, and economics.[38]

Webster's criticism prompted soul-searching by other researchers. In her 1997 Wood-Gush Memorial Lecture, Marian Stamp Dawkins responded by asking why there had not been more progress in animal welfare research. According to Dawkins, the last 20 years had in fact seen significant progress, but many initial assumptions had proven too simplistic: research had shown that measuring animal consciousness, cognition, and emotions was more complex than expected; it had become clear that behavioural differences between wild and captive animals and 'vacuum' or stereotyped behaviours were not necessarily indicative of suffering; and measuring hormonal stress indicators without simultaneously assessing animals' experiences had proven misguided: "We must not over-simplify that which is complicated."[39] This did not mean that research was worthless or not "pragmatic, utilitarian, and circumspect"[40] enough to produce good standards. Acknowledging complexity was a precondition for meaningful change. This was particularly true regarding animal feelings.

[34] Webster, *Cool Eye*, 130.
[35] Webster, *Cool Eye*, 240; see also 10.
[36] Webster, *Cool Eye*, 241.
[37] Webster, *Cool Eye*, 251.
[38] Webster, *Cool Eye*, 259.
[39] Stamp Dawkins, "Why has there not been more progress," 66; see also: Stamp Dawkins, "From an animal's point of view," 4–5.
[40] Stamp Dawkins, "Why has there not been more progress," 66.

In her influential 1990 essay "From an animal's point of view," Dawkins had already argued that engaging with law-making would require scientists to enter the "muddy waters"[41] of studying positive and negative feelings in more detail. It was easy for activists, politicians, and philosophers to take the "moral high ground" and drive changes in law based on "gut feelings"[42] about animal preferences. However, resulting laws and standards risked being ineffective. Integrating physiological, ecological, and affective approaches was the best way to develop meaningful welfare guidelines. In the case of animal suffering, Dawkins argued for an inclusive assessment of the 'canonical costs' to an animal of preserving its fitness and the 'perceived costs' by the animal itself—even if there might be no threat to its physical welfare.[43] This required a holistic ethological approach: "You have to be an ethologist as Tinbergen conceived one—that is, to understand, amongst other things, what the animal's natural behaviour is, what it is adapted to, how it acquires the relevant information as well as how it acquires and processes sensory information."[44]

For Dawkins, this meant applying complex economics of choice based on comparative demand and income curves to test animal preferences.[45] Other researchers continued to favour different approaches. In Cambridge, Donald Broom warned against prioritising preference over "direct measures of welfare":

> The term 'welfare' should refer to a characteristic of an individual at the time under consideration, that is, to its state rather than to anything which is given to that individual. When conditions are favourable, animals regulate their interactions with their environment without difficulty. Under hostile conditions, animals use various methods to try to counteract the adverse effects of those conditions. These attempts to cope can themselves be measured and, if they fail, adverse effects on the animal can be measured. The welfare of an individual is the state resulting from its attempts to cope with its environment.[46]

[41] Stamp Dawkins, "From an animal's point of view," 1.
[42] Stamp Dawkins, "Why has there not been more progress," 67.
[43] Stamp Dawkins, "From an animal's point of view," 3.
[44] Stamp Dawkins, "Why has there not been more progress," 72.
[45] Stamp Dawkins, "From an animal's point of view," 1–9.
[46] Donald M. Broom, "The importance of measures of poor welfare – response to Marian Dawkins 'From an Animal's Point of View'," *Behavioural and Brain Sciences* 13 (1990), 14.

Welfare could be measured on a scale from very good to very poor: if an individual failed to cope with an environment, its life would be adversely affected and welfare was poor; if it coped but with great difficulty, welfare would also be poor. Defending himself against accusations of over-emphasising physiological measurements,[47] Broom noted that research on animal feelings and preferences provided "valuable indirect measures"[48] for welfare but had to be contextualised—sometimes animals chose situations that were demonstrably bad for them: "Welfare cannot be assessed by preference studies alone, however; veterinary surgeons' vast knowledge concerning the recognition of signs of injury or ill health and the rapidly increasing number of other indicators of poor welfare must be used, too."[49]

Sussex-educated US researcher Joy Mench was sceptical of both approaches. In a 1998 paper for *Applied Animal Welfare Science*, she noted: "There is (…) a growing sense that animal welfare science has reached an impasse and that ethical and scientific questions (…) have become hopelessly entangled."[50] Overcoming this impasse would depend on moving beyond the post-Brambell focus on suffering and preference indications to "broader quality-of-life questions."[51] The 1965 report had stimulated a productive emphasis on the minimisation of pain and suffering. However, this approach was running out of steam. Producers were only slowly adopting new housing systems, and welfare as a definitive concept remained elusive:

> Behavioral and physiological measures both have important limitations, may be inconsistent with each other, and can be difficult to interpret because their expression is influenced by many complex factors including individual predispositions. Perhaps more important, there seems no clear way to establish a cutoff point below which welfare is 'bad.'[52]

[47] Broom, "A History of Animal Welfare Science," 127.

[48] Broom, "The importance of measures of poor welfare," 14.

[49] Broom, "The importance of measures of poor welfare," 14; see also Broom's 1998 reply to Mench (discussion below) in which he emphasized the need to measure positive aspects of welfare to balance the field, Donald M. Broom, "Welfare as a Broad Scientific Concept," *Journal of Applied Animal Welfare Science* 1/2 (1998), 149–151.

[50] Joy A. Mench, "Thirty Years After Brambell: Whither Animal Welfare Science?," *Journal of Applied Animal Welfare Science* 1/2 (1998), 91.

[51] Mench, "Thirty Years After Brambell," 91.

[52] Mench, "Thirty Years After Brambell," 92.

The lack of a welfare cutoff and measurement disagreements had resulted in a situation where welfare was defined in minimalist terms to gain agreement on basic principles. Linking welfare to the absence of suffering had also led to the relative neglect of positive feelings and of animal behaviour that could not be linked to suffering.

According to Mench, it was time for scientists "to make an ethical leap"[53] and formulate a broader operational definition of animal welfare that incorporated a high level of biological functioning, freedom from suffering, and positive feelings. Taking a "quality of life definition" would help overcome false dichotomies like good and bad welfare, distress and eustress, and luxuries and necessities, because "welfare will depend on the relative preponderance of positive over negative experiences during the animal's lifetime."[54] Established deprivation experiments where the effects of stripping back 'amenities' could be measured were perfectly suited to assessing positive feelings.[55] Findings could be used for an "additive model"[56] of welfare where one could start enriching existing intensive environments.

Concerned about the effects of internal disagreements on their discipline, Ian Duncan and David Fraser called on welfare researchers to devote less attention to abstract 'measuring' debates and concentrate on identifying and solving concrete welfare problems. Many problems like hunger or distress were obvious.[57] Others entailed a more detailed study of animals' physiology, preferences, affective states, and the adaptive value of specific behaviour.[58] Open research questions were, however, no excuse for inaction.

It was also clear that scientists could do more to engage other fields. In his 1999 Wood-Gush Memorial Lecture, Fraser criticised the fact that welfare researchers had been remarkably "selective in acknowledging the role of these ethicists and critics."[59] Lack of scientific engagement had been

[53] Mench, "Thirty Years After Brambell," 94.

[54] Mench, "Thirty Years After Brambell," 97.

[55] Stolba and Wood-Gush, "The identification of behavioural key features," 287–298.

[56] Mench, "Thirty Years After Brambell," 98.

[57] Ian J.H. Duncan, "Thirty Years of Progress in Animal Welfare Science," *Journal of Applied Animal Welfare Science* 1/2 (1998), 152–153.

[58] Fraser and Duncan, "'Pleasures', Pains' and Animal Welfare," 383–396.

[59] David Fraser, "Animal ethics and animal welfare science: bridging the two cultures – the D.G.M. Wood-Gush Memorial Lecture," *Applied Animal Behaviour Science* 65 (1999), 173.

"fully reciprocated"[60] by thinkers like Regan and Singer, who prioritised justice over caring for animals. Fraser instead engaged with philosopher Bernard Rollin's criticism of positivist welfare research.[61] Scientists and ethicists alike should avoid a priori exclusions of each other's approaches.[62] Scientific debates about whether welfare centred on survival, health, and comfort or whether it should also encompass sentience were mirrored in philosophical debates:

> Most attempts by scientists to conceptualize and study animal welfare boil down to three key issues: that animals should feel well (...) that animals should function well (...) and that animals should lead natural lives (...). These ideas correspond at least roughly to the concepts of 'interests', 'needs', and telos, respectively as defined by some philosophers.[63]

Scientists needed to conduct and contextualise their work within broader care-based ethical frameworks and abandon the idea that research was strictly objective. Ethicists needed to be more specific about the empirical foundations of their frameworks:

> I think we can view animal welfare as an evaluative concept (...). Animal welfare encompasses many variables that can be studied scientifically and objectively. However, our decisions about which variables to study, and how to interpret them in terms of an animal's welfare, involve normative judgements about what we consider better or worse for the quality of life of animals.[64]

Fraser's call for methodological and interdisciplinary openness proved prescient. While welfare researchers had managed to attain unprecedented influence in academia, politics, and industry, epistemic disagreements threatened to stall the discipline's momentum. It was by distancing themselves from overly positivist approaches and openly acknowledging their

[60] Fraser, "Animal ethics and animal welfare science," 173; see also 175.

[61] Bernard E. Rollin, "Animal Production and the new social ethic for animals," *Journal of Social Philosophy* 25th special issue (1994), 71–83.

[62] Fraser, "Animal ethics and animal welfare science," 177.

[63] Fraser, "Animal ethics and animal welfare science," 178.

[64] Fraser, "Animal ethics and animal welfare science," 193; the need to reengage philosophers and ethicists had already been highlighted by University of Colorado ethologist Marc Bekoff in 1991; Marc Bekoff, "The animal's point of view, animal welfare and some other related matters," *Behavioral and Brain Sciences* 14/4 (1991), 753–755.

status as practitioners of a value-influenced science that welfare researchers have productively engaged the described *Sinnkrise* of their discipline.

Ironically, part of this process has entailed a return to the explicitly normative considerations guiding early ethologists like Julian Huxley and William Thorpe in their evaluation of animal sentience and humans' obligations towards animals. This does not mean that research has 'regressed' since the 1990s. Their (re-)engagement with ethical considerations instead signals that welfare researchers no longer see themselves as comprising a sub-field that must avoid accusations of anthropomorphism at all costs but rather as a self-confident discipline capable of embracing complexity and avoiding overhasty promises of universal standards.

With senior researchers like Fraser, Webster, Duncan, Broom, and Dawkins now in their 60s, 70s, and 80s, a new generation of welfare scientists has begun to ask different questions and apply new techniques ranging from genomics and epigenetics to machine-learning in order to inform evidence-based veterinary medicine and farm design.[65] This does not mean that value debates and questions of how to deal with politicians and market actors have disappeared. Edinburgh researcher David Mellor has recently called for a 'positive' reframing of the original Five Freedoms as five welfare provisions with aligned welfare aims. The aim is to avoid popular yet unhelpful conflations of freedoms with rights and to stop primarily defining well-being as the absence of negative experiences.[66] There are also ongoing debates on non-stun slaughter, the ethical trade-offs involved in reducing antibiotic use, and the use of constructive anthropomorphism in qualitative behaviour assessment.[67] Agreeing on the relative weighting of different welfare indicators also remains challenging. In 2017, the UFAW organised a symposium titled "Measuring animal welfare and applying scientific advances—why is it still so difficult?" Questions

[65] For an overview of current techniques see: Michael C. Appleby, Anna Olsson and Francisco Galindo (eds.), *Animal welfare* (Wallingford and Oxford: CABI, 2018).

[66] David J. Mellor, "Moving beyond the 'Five Freedoms' by Updating the 'Five Provisions' and Introducing Aligned 'Animal Welfare Aims'," *Animals* 6/10 (2016), 59.

[67] M. Haluk Anil, "Religious slaughter: A current controversial animal welfare issue," *Animal Frontiers* 2/3 (2012), 64–67; Alexander Trees, "Non-stun slaughter: the elephant in the room," *Veterinary Record* 182/7 (2018), 177; Richard Helliwell, Carol Morris, and Sujatha Raman, "Antibiotic stewardship and its implications for agricultural animal-human relationships: Insights from an intensive dairy farm in England," *Journal of Rural Studies* 78 (2020), 447–456; Michal Arbilly and Arnon Lotem. "Constructive anthropomorphism: a functional evolutionary approach to the study of human-like cognitive mechanisms in animals," *Proceedings of the Royal Society B: Biological Sciences* 284/1865 (2017), 20171616.

identified were "Will we ever be able to demonstrate sentience? (…) Are the techniques that we have to study emotional state (affect) adequate (…)? How important is positive welfare? (…)? How robust is the data collected on animal welfare?"[68]

However, these discussions no longer seem to pose a wider *Sinnkrise* for the discipline. Reviving Fraser's exhortations for scientists to engage other research traditions,[69] a 2014 FAWC review noted that farm animal welfare was about more than animals. Welfare and welfare politics remained inextricably affected by wider cultural and socio-economic values: "The key issue is that there is no gold standard for animal welfare, i.e. no one absolute measure that always and only identified that an animal has poor or good welfare. (…). Results and observations are interpreted by humans and accepted by some and not others."[70] Although hopes for bias-free universal welfare indicators remain, the normative considerations that triggered Ruth Harrison's initial turn towards animal welfare in 1961 are unlikely to ever disappear from our thinking, research, and treatment of animals.

[68] Stephen Wickens, Robert Hubrecht and Huw Golledge, "Welcome to the UFAW Symposium," *Conference Program: Measuring Animal Welfare and Applying Scientific Advances: why is it still so difficult?* 27th-29th June 2017, Royal Holloway, University of London, Surrey, UK; https://www.ufaw.org.uk/downloads/ufaw-symposium-royalh-2017%2D%2D-conference-booklet-v3-online.pdf [13.04.2020]; see also similar questions in FAWC, *Evidence And The Welfare Of Farmed Animals. Part 1: The Evidence Base* (London: FAWC, 2014), 3.

[69] FAWC, *Evidence and the Welfare of Farmed Animals*, 19 (footnote 16).

[70] FAWC, *Evidence and the Welfare of Farmed Animals*, 34; Buller and Roe, *Food and Animal Welfare*, 33–41.

CHAPTER 13

Conclusion

Ruth Harrison would not have been surprised by the enduring influence of value-based judgement within animal welfare science. Rooted in the synthesist principles of Edwardian reform, she would have argued that only a value-based science could inform the humane treatment of animals in a morally progressive society. There was no contradiction between science, activism, and politics. However, she did not live to see the most recent resurgence of value debates among a new generation of welfare scientists.

Approaching her 80th birthday, Harrison and other long-standing FACT members like David Sainsbury and Andrew Fraser resigned from the organisation in September 1999.[1] Harrison had created FACT and steered its development more or less single-handedly for over 32 years. With FACT chairmanship passing to Donald Broom—and later Marian Stamp Dawkins[2]—it was clear that "FACT would be entering a new era" and would have to "stand on its own two feet."[3] Subsequent restructuring

[1] FACT Files, MD, Minute Book, Farm Animal Care Trust, Minutes of a Meeting of Trustees (13.09.1999).

[2] FACT Files, MD, Minute Book, Farm Animal Care Trust, Meeting of Trustees, Minutes of meeting (11.05.2000).

[3] FACT Files, MD, Minute Book, Farm Animal Care Trust, Meeting of Trustees (11.08.2000).

C. Kirchhelle, *Bearing Witness*, Palgrave Studies in the History of Social Movements,
https://doi.org/10.1007/978-3-030-62792-8_13

occurred without input from Harrison, who died of cancer in June 2000.[4] Obituaries praised Harrison's tenacity and impact on animal welfare politics but also noted her chronic dissatisfaction with progress.[5] While Harrison's long-term project of writing a second *Animal Machines* remained unfinished,[6] Carol Mckenna listed some of her most important achievements in the *Guardian's* obituary:

> In her lifetime she saw many improvements. Veal crates (1990) and sow/tether stalls (1999) become illegal in Britain. Last year saw the announcement that battery cages will be phased out by 2012.[7]

Meeting two months after her death, FACT trustees noted:

> Ruth had been probably the most important and influential single person in the early recognition of the threat to animal welfare inherent in many modern intensive farming methods, and a prime mover in the emergence and development of the scientific investigation of welfare in farm animals.[8]

However, as years passed, a chronological shortening of Harrison's campaigning biography set in. Fellow activists, animal welfare researchers, and historians glossed over her 36 years of full-time campaigning, 32 years of research sponsorship via FACT, 24 years on FAWAC and FAWC, and 6 turbulent years on the RSPCA Council. Harrison's impact was thus increasingly equated with her book. Within the welfare community, she was portrayed as an iconic yet chronologically distant Carson-like founding figure. In 2013, the University of Oxford organised a conference to highlight the achievements of Ruth Harrison and to celebrate the reprint of *Animal Machines.* In his foreword to the new edition, John Webster noted:

> Today, *Animal Machines*, the book, should be read the way one reads Aristotle or the Bible: with great respect for its power and insight, but not

[4] FACT Files, MD, Minute Book, Farm Animal Care Trust, Meeting of Trustees (11.08.2000).

[5] Oral History Interview Ruth Layton (02.07.2014).

[6] Correspondence with Marlene Halverson January–February 2014.

[7] Carol McKenna, "Ruth Harrison", *Guardian*, 06.07.2000, 22.

[8] FACT Files, MD, Minute Book, Farm Animal Care Trust, Meeting of Trustees (11.08.2000).

> to be taken as gospel. Much of what she describes has changed, … Nevertheless, the evolution of major improvements in farm animal welfare for pigs, calves and chickens through legislation in the UK and European Union, the state-by-state legislation to ban sow stalls in the USA, the development of high welfare schemes like Freedom Foods and the Global Animal Partnership, and the massive increase in funding for the pursuit and application of animal welfare science … can all be traced back, like mitochondrial DNA (female line), to the common ancestor, namely Ruth herself.[9]

As this book has shown, looking not just at the author but at the person Ruth Harrison reveals a much more multifaceted story of generational change and dynamic interactions between animal welfare politics, activism, and science. During her life, synthesist Edwardian campaigning gave rise to professionalised activism and new concepts of animal cognition, affective states, and welfare. The backstage of British corporatist welfare politics was similarly transformed by polarising frontstage public protest and radical animal rights thinking. Aided by the rise of a new mandated form of animal welfare science and European integration, the turbulent 1970s eventually resulted in a new era of British welfare politics characterised by transnational decision-making and market-driven assurance schemes, which relied on consumer citizens rather than citizen campaigners to drive change.

Reinserting the person Ruth Harrison back into this networked world and using her biography to study it reveals these wider dynamics of twentieth-century animal welfare. It may also debunk some of the hagiography, which has risen around her, but does not diminish her achievements. Ruth Harrison was clearly not the overly sentimental, timid, or conservative housewife that critics made her out to be. Neither was she a one-hit author, who came from and vanished into nowhere. Instead, she was a well-educated, well-connected successful campaigner, who was shaped by the synthesist vegetarian and pacifist values of Edwardian reform and whose defining characteristic was the "relentless vigour"[10] with which she campaigned against the inhumane treatment of humans and non-humans.

This relentlessness was already evident when Harrison interrupted her education to work as an FAU nurse in bombed-out British cities and as a

[9] Webster, "Ruth Harrison – Tribute to an Inspirational Friend," 6.
[10] Webster, "Ruth Harrison – Tribute to an Inspirational Friend," 8.

relief worker in post-war Germany. As a convinced Quaker, Harrison believed in living faith through action and non-violent change by bearing witness against grievances. Similar to many other Quakers, she was attracted by the new forms of civic protest inaugurated by the CND in post-war London and shared popular contemporary concerns about the detrimental effects of technological development on the environment, health, and social ethics. Harrison's vision of broader moral, environmental, and societal reform was shared by many other contemporaries including leading British animal researchers like William Homan Thorpe and Julian Huxley. Disagreeing with mechanistic behaviourist models and continental ethologists' decision to shy away from affective states, they saw the study of animal consciousness and cognitive evolution as key to developing a progressivist post-war programme of social and moral reform. Their ambition opened the door for the scientific acknowledgement of animal feelings beyond pain and also entailed seeing the humane treatment of all animals as a prerequisite for social and scientific progress.

By 1960, this vision of humane social reform was seemingly threatened by the dystopian "sociotechnical imaginary"[11] of the factory farm. Concerns about intensive livestock operations' health, environmental, and moral impacts on the self-described 'Nation of Animal Lovers' created a fertile meeting ground for scientists and activists. Ruth Harrison's talent as an author lay not in being the first to identify and target this meeting ground but in successfully staging its underlying dystopian imaginary for wider audiences. After failing to convince Britain's Society of Friends to join her campaign, she spent the years between 1961 and 1964 scouring relevant literature, contacting various political and activist organisations, and writing her future bestseller. An especially fruitful result of Harrison's networking was her contact with US environmentalist Rachel Carson. The correspondence between the two iconic authors reveals how closely post-war environmentalism and animal activism were entwined. On both sides of the Atlantic, leading campaigners came from similar backgrounds of radical reform and synthesist progressivism and shared a basic set of environmental and moral concerns about technology's impacts on humanity. Appearing within two years of each other, *Silent Spring* and *Animal Machines* contained similar core messages and helped turn intensive agriculture and associated technologies like DDT into new focal points for contemporary protest movements.

[11] Jasanoff and Sang-Hyun, "Sociotechnical Imaginaries," 189–196.

Aided by a skilful promotion campaign in the *Observer*, *Animal Machines'* bestselling success and resulting public outrage led to the installation of the Brambell Committee. The committee's pioneering 1965 report combined existing concepts of cruelty with new behavioural welfare considerations. It also recommended legislative reform alongside a new permanent welfare body to evaluate and guide British policy. Despite her role in triggering the installation of the Brambell Committee, Harrison was only invited to provide evidence. The realisation that resting on her bestseller laurels would not allow her to influence welfare reform made her decide to become a full-time activist. Between 1966 and 1969, Harrison used her status as a non-aligned yet widely trusted outsider to relentlessly lobby for a nomination to the government's new FAWAC, founded her own research trust, and was elected onto the RSPCA Council.

On FAWAC, Harrison did not turn out to be the 'easy' choice envisioned by MAFF officials. Faced with a pro-industry majority on the committee, Harrison and other welfarists adopted a dual strategy of blocking weak compromises while simultaneously applying external pressure to push for improved codes. This dual strategy could prove remarkably successful—as in the case of the 1969 welfare code revisions. However, it also contributed to a breakdown of FAWAC decision-making, a resulting lack of meaningful code reforms, and a fraying of formerly consensus-oriented corporatist welfare decision-making.

The stagnation of backstage welfare reform contributed to the 1970s polarisation of public frontstage animal welfare politics. Ruth Harrison struggled to navigate this increasingly crowded political marketplace. Within the RSPCA, her decision to leak the BFSS letter ended her short but fruitful alliance with RSPCA traditionalists. However, her ongoing FAWAC membership and focus on gradual welfare improvements also made it difficult to form new alliances with younger, more radical campaigners in the Reform Group, who viewed older female campaigners like Harrison as being too timid to stand up for animal interests. In contrast to Harrison's contractualist understanding of humans' duty towards fellow creatures, younger activists employed concepts of speciesism and animal rights to oppose intensive animal husbandry per se.

Although Harrison exited the RSPCA Council only in 1975, she found herself isolated in an organisation that was rapidly changing in response to growing demands on its organisational capabilities and the end of post-war establishment politics. There was increasingly little space for self-described 'loners' like Harrison in this new corporatist world of professionalised campaigning. Between 1970 and 1974, the still 'traditionalist' Council had formed new expert advisory committees and launched successful media campaigns against live animal exports. The 1974 inquiry and 1977 election of Richard Ryder as chairman ended long-standing internal tensions over hunting and resulted in a further streamlining of management, opening of leadership structures, and focus on animal rights.

Culminating in a lost libel case and personal bankruptcy, Harrison's six years on the RSPCA Council soured future relations with the Society. However, her experience within FAWAC and the RSPCA also made her realise the growing importance of mobilising scientific support and data for welfare campaigning. During increasingly charged discussions on new welfare codes and regulations, relying solely on ethical or moral argumentation proved insufficient to counter industry arguments that existing practices did not harm animals. Alongside the RSPCA's Farm Livestock Advisory Committee (FLAC), Harrison began to intensify relations to the new discipline of farm animal welfare science and used FACT to fund supportive research.

For animal welfare scientists, resulting sponsorship was both a chance and a challenge. After 1965, welfare research had initially been dominated by veterinary scientists, who were intent on defining physiological indicators of inadequate welfare. During the 1970s, that early emphasis on pain and stress was supplemented with a new behavioural focus on 'abnormal' farm animal behaviour, 'natural' husbandry environments, and animal preferences. With classic ethology beginning to fragment, younger researchers were attracted to farm animal welfare because of the possibility it offered to conduct and apply behavioural research. The applied aspect of their research allowed welfare researchers to tap into new governmental and non-governmental funding stream for outcome-oriented research. While MAFF and FAWAC initially prioritised physiological research on pain and productivity, behavioural researchers profited from anti-vivisectionist taboos by welfarist sponsors like the RSPCA. Similar to synthesist 1950s ethology, resulting research protocols were a chimera of hypothesis-driven science, economic interests, and value-based debates on animals' place in society and the meaning of welfare.

Rising funding supported an institutionalisation and expansion of British animal welfare science. However, the field's status as a mandated science in a polarised environment also meant that researchers had to be politically circumspect. As RSPCA conflicts over the meaning of animal experimentation and Harrison's attempts to influence research show, scientists had to balance funder expectations for useful results with the need to maintain authority over research protocols. Maintaining expert authority was further complicated by a lack of consensus over welfare definitions. While scientists agreed that welfare could be measured, initial hopes for universal welfare indicators proved premature, and it remained unclear how behavioural and physiological research results could be combined.

Clashes over whose authority to trust with regard to animal welfare reached a climax in 1979 when the RSPCA disbanded FLAC and boycotted the new FAWC. The move highlighted deep rifts between moderate and more radical activists over whether to continue cooperating with official bodies and how far to trust scientists as arbiters of animal welfare rather than rights. In 1980, the reversal of the RSPCA boycott marked a significant victory of moderates.

The episode also revealed how polarised and dysfunctional British farm animal welfare politics had become after a decade of relative neglect in Downing Street. Following heightened activity between 1964 and 1970, FAWAC's breakdown and MAFF inaction had led to a relative stagnation of British farm animal welfare reform. Political momentum for further reforms now frequently came from the continent. Countries like West Germany passed more stringent legislation, and major welfare decisions were increasingly made at the European level in the wake of Britain's 1973 EEC accession and the 1976 Council of Europe Convention for the Protection of Animals Kept for Farming Purposes.

Reacting to the growing popularity of welfare issues among voters in 1979, the new Thatcher government was not only keen to highlight its welfare credentials but also less committed to maintaining the traditional authority of MAFF and producer organisations over welfare politics. The result was both a reinvigoration of British farm animal welfare politics and a gradual shift towards market-driven standard-setting. Officially established in July 1979, the new FAWC was given independence from MAFF, staffed with more welfare scientists, and allowed to explicitly reference an expanded version of the five freedoms in its brief. The committee's rapid revision of welfare codes and recommendation of positive welfare changes ended FAWAC's regulatory deadlock and boosted the status of welfare researchers and FAWC welfarists like Ruth Harrison. Now in her 60s, the veteran campaigner witnessed the fulfilment of key demands from *Animal Machines* such as Britain's effective abolishment of intensive veal husbandry. Harrison's rising social standing also enabled her to increase

FACT resources for targeted welfare research and political networking at the British and European level. This influence only gradually diminished after Harrison's withdrawal from FAWC and the European T-AP during the 1990s. By the time of her death in 2000, Harrison was widely recognised as a thorny yet respectable establishment spokesperson for animal welfare.

The political economy of farm animal welfare had also changed. During the first decades after 1945, British welfare politics had been dominated by MAFF and industry-weighted corporatist advisory committees consisting of hand-picked welfarist and industry representatives. Following the effective breakdown of this system, the 1980s and 1990s saw an increasingly powerful second tier of privatised welfare politics emerge. Official bodies like FAWC and T-AP continued to play an important role in setting minimum standards. However, the increasing segmentation of the food market also created lucrative premium niches for products whose ethical and health properties had to be certified. Following the lead of the organic sector, animal welfare scientists, charities, and supermarkets formed powerful welfare assurance schemes. The so-called virtuous bicycle of welfare schemes was driven both by consumers' desire to pay for the ethical production of animals and by the increasing hold of a few retailers and integrated companies over British agricultural politics and farmers. Non-statutory assurance schemes also minimised public conflicts over welfare by delegating discussions over standard-setting to a more difficult-to-access corporate and expert-dominated 'backstage'.

Despite profiting from the new revenue streams unleashed by the new assurance schemes, scientists continued to disagree about the weighting of different welfare indicators. However, the resulting *Sinnkrise* was only temporary. Welfare scientists' growing engagement with ethicists and social scientists is indicative not only of the limits of a purely positivist approach that focuses on defining and measuring universal welfare parameters but also of a growing acknowledgement of the value-based side of welfare in a less polarised political environment.

Ruth Harrison would have certainly endorsed welfare scientists' reengagement with value debates. She may, however, have been more sceptical about the increasing status of animal welfare as an economic value. In 1964, both Harrison and Carson called for a consumers' revolt against intensive farming. Over half a century later, British animal production had indeed experienced a demand-led shift towards enhanced welfare. This shift was in part based on an increasingly robust British and European regulatory framework and in part on voluntary self-regulation via assurance schemes. Millions of farm animals now live in scientifically vetted, welfare-conducive environments. It is, however, doubtful whether the chronically dissatisfied Harrison would have been content with a situation in which welfare functions as commodity that is selectively applied to add value to certain segments of animal production and not to others. Although animal welfare is now firmly established as part of mainstream politics and agrocapitalism, it has to a certain extent been divorced from the universalist moral framework that Ruth Harrison decided to bear witness to.

Bibliography

Archives

Bayerisches Hauptstaatsarchiv [Bavarian Main State Archive], Munich, Germany (HSTA)
British Library (Ryder Papers), London, UK
Cambridge University Library (Thorpe Papers), Cambridge, UK
FACT source material held by Prof Marian Stamp Dawkins (MD), Oxford, UK
FACT source material held by Prof Donald Broom (DB), Cambridge, UK
Library of the Society of Friends (FAU Papers), London, UK
Rice University (Julian Sorrell Huxley Papers), Houston, USA
RSPCA Archives, Horsham/ Southwater, UK
The British National Archives, Richmond, UK (TNA)
Whitechapel Gallery Archive, Clare Winsten Autobiography, London, UK
Yale Beinecke Library, (Carson Papers), New Haven, US (YBL)

Oral History Interviews & Correspondence

Sir David Attenborough (19.08.2015)
Prof Donald Broom (04.07.2014)
Prof Marian Stamp Dawkins (01.07.2014; 07.08.2015)
Jonathan Harrison (21.01.2015; 29.08.2015)
Dr Ruth Layton (02.07.2014)
Dr Richard Ryder (18.01.2015; 10.08.2015; 23.08.2015)
Prof Peter Singer (17.01.2015)

C. Kirchhelle, *Bearing Witness*, Palgrave Studies in the History of Social Movements,
https://doi.org/10.1007/978-3-030-62792-8

Serial Sources

British Farmer
Daily Mail
Farmers Weekly
Glasgow Herald
Guardian
Jewish Telegraph
New York Times
Observer
Reading Eagle
The London Gazette
The Times
Washington Post

Digital Sources

"1979 Conservative Party General Election Manifesto", http://www.conservativemanifesto.com/1979/1979-conservative-manifesto.shtml [02.02.2021].

"1979 Labour Party Manifesto. The Labour Way is the Better Way", http://www.labour-party.org.uk/manifestos/1979/1979-labour-manifesto.shtml [01.09.2020].

"1997 Labour Party Manifesto. New Labour Because Britain Deserves Better", http://www.labour-party.org.uk/manifestos/1997/1997-labour-manifesto.shtml [15.04.2020].

"1997 Conservate Party Manifesto. You can only be sure with the Conservatives", http://www.conservative-party.net/manifestos/1997/1997-conservative-manifesto.shtml [12.12.2020].

"2013 Prof John Webster and Prof Peter Sandøe", *UFAW Medal for Outstanding Contributions to Animal Welfare Science*, https://www.ufaw.org.uk/ufaw-medal-for-outstanding-contributions-to-animal-welfare-science/ufaw-medal-for-outstanding-contributions-to-animal-welfare-science-past-awards#webster [10.04.2020].

Agriculture (Miscellaneous Provisions) Act 1968, Legislation.gov.uk; URL: http://www.legislation.gov.uk/ukpga/1968/34 [09.01.2015].

Andrew Fraser, "A Short Biography of Andrew Fraser, written by him in March 2008a", *Applied Ethology.org*, https://www.applied-ethology.org/res/dr_%20andrew%20fraser_%20isae%20honorary%20fellow.pdf [01.05.2020].

Ayad, Sara, "The Winstens of Whitechapel: Clara Birnberg and Simy Weinstein," *Art UK*, https://artuk.org/discover/stories/the-winstens-of-whitechapel-clara-birnberg-and-simy-weinstein [22.02.2021].

Birnberg, Ariadne, *Most Beautiful Maynard*, https://longandvariable.files.wordpress.com/2015/05/most-beautiful-maynard.pdf [01.05.2020].

Churchill, Winston, "Their Finest Hour", 18.06.1940, House of Commons, https://winstonchurchill.org/resources/speeches/1940-the-finest-hour/their-finest-hour/ [13.05.2020].

"Chapter 11: Membership", *Quaker Faith and Practice* 5th edition, https://qfp.quaker.org.uk/chapter/11/ [08.11.2019].

"Dex Harrison – Basic Biographical Details", *Dictionary of Scottish Architects, Architect Biography Report*: http://www.scottisharchitects.org.uk/architect_full.php?id=206027 [20.12.2014].

"Ethology: Claims and Limits of a lost Discipline", Podcast Series Wissenschafts Portal Gerda Henkel Stiftung, https://lisa.gerda-henkel-stiftung.de/ethology_claims_and_limits_of_a_lost_discipline_podcast_series?nav_id=9149 [15.05.2021].

"Europe: Meat Output Statistics", in *International Historical Statistics*, (London: Palgrave Macmillan, April 2013).

European Convention for the Protection of Animals kept for Farming Purposes (Strasbourg, 10.03.1976): URL: http://conventions.coe.int/Treaty/EN/Treaties/Html/087.htm [17.12.2014].

"Farm Animal Care Trust": http://www.fact.uk.com [09.01.2015].

"Foxhunting Protest At RSPCA Meeting", *ITN Source. JISC MediaHub:* http://jiscmediahub.ac.uk/record/display/039-00043880;jsessionid=5163BD6FF3274A40C42880EFD9825FF9 [01.12.2014].

House of Commons, Early Day Motion No. 1543, "Freedom Food", Tabled 19.07.1994, 1993–1994 Session, https://edm.parliament.uk/early-day-motion/8714/freedom-food [01.05.2020].

Margaret Thatcher Foundation, "Shadow Cabinet: Circulated Paper, The Conservative Manifesto 1978 - The Right Approach to Government. 2nd LCC Draft. Copy No. 6", 18–19, URL: https://c59574e9047e61130f13-3f71d0fe2b653c4f00f32175760e96e7.ssl.cf1.rackcdn.com/8C1B6421465247B4BB8F6BE90097961B.pdf [01.04.2020].

"Quakers in Action. FAU in WWII: Civilian Relief Work in Mainland Europe", *Quakers in the World*: http://www.quakersintheworld.org/quakers-in-action/295, [08.06.2016].

"Quakers in Action. Women in the FAU", *Quakers in the World*: http://www.quakersintheworld.org/quakers-in-action/329, [08.06.2016].

"Stephen Winsten, 1893–1991", *Remembering the men who said no, conscientious objectors 1916–1919, Peace Pledge Union project*, URL: https://menwhosaidno.org/men/men_files/w/winstent_s.html [30.04.2020].

"The Nobel Prize in Physiology or Medicine 1973", *The Nobel Prize*, https://www.nobelprize.org/prizes/medicine/1973/summary/ [20.05.2020].

"UFAW Medal for Outstanding Contributions to Animal Welfare Science - past awards", *Universities Federation for Animal Welfare,* URL: https://www.ufaw.org.uk/ufaw-medal-for-outstanding-contributions-to-animal-welfare-science/ufaw-medal-for-outstanding-contributions-to-animal-welfare-science-past-awards [13.04.2020].

Web of Science, "Web of Knowledge" search of term "animal welfare" [05.04.2020].

Wickens, Stephen, Robert Hubrecht and Huw Golledge, "Welcome to the UFAW Symposium", *Conference Program: Measuring Animal Welfare and Applying Scientific Advances: why is it still so difficult? 27th–29th June 2017, Royal Holloway, University of London, Surrey, UK*; URL: https://www.ufaw.org.uk/downloads/ufaw-symposium-royalh-2017%2D%2D-conference-booklet-v3-online.pdf [13.04.2020].

Wild, Simon, "Henry S. Salt", *Henry S. Salt Society*, https://www.henrysalt.co.uk/life/biography/ [01.05.2020].

Printed Material

Abraham, John, *Science, Politics and the Pharmaceutical Industry. Controversy and bias in drug regulation* (London and New York: Routledge, 1995).

Agar, Jon, *Science in the 20th Century and Beyond* (Cambridge: Polity, 2012).

Agar, Jon, *Science Policy Under Thatcher* (London: UCL Press, 2019).

Agricultural Research: Background and Issues (Washington DC: Congressional Research Service, 2020).

Aiston, Sarah, "A Good Job for a Girl? The Career Biographies of Women Graduates of the University of Liverpool Post-1945," *Twentieth Century British History* 15/4 (2004), 361–387.

Anil, M. Haluk, "Religious slaughter: A current controversial animal welfare issue," *Animal Frontiers* 2/3 (2012), 64–67.

"Animals (Control of Intensified Methods of Food Production)," *Hansard* Vol. 630 (23.11.1960).

Anon., "Birnberg, Benedict Michael," in W. Rubinstein and Michael Jolles (eds), *Palgrave Dictionary of Anglo-Jewish History.*

Anon., "Winsten, Clare & Stephen," in W. Rubenstein and Michael Jolles (eds), *Palgrave Dictionary of Anglo-Jewish History.*

Anon, "Domesticated animals breed on regardless," *New Scientist* (14.06.1973), 665.

Anon, "Is Ethology Respectable?," *Nature* 216/5110 (1967), 10.

Anon, "Society For Veterinary Ethology, "'Stress in Farm Animals' – proceedings of joint symposium with the Royal Society for the Prevention of Cruelty to Animals, London 25–26, May 1973," *British Veterinary Journal* 130 (1974), 85–95.

Appleby, Michael C., Anna Olsson, Francisco Galindo (eds)., *Animal welfare* (Wallingford and Oxford: CABI, 2018).

Arbilly, Michal, and Arnon Lotem. "Constructive anthropomorphism: a functional evolutionary approach to the study of human-like cognitive mechanisms in animals," *Proceedings of the Royal Society B: Biological Sciences* 284/1865 (2017): 20171616.

Barad, Karen, *Meeting the Universe Halfway: Quantum Physics and the Entanglement of Matter and Meaning* (Durham, NC: Duke University Press, 2007).

Bashford, Alison, "Julian Huxley's Transhumanism," in Marius Turda (ed.), *Crafting Humans: From Genesis to Eugenics and Beyond* (Göttingen and Taipei, V&R Uni Press/National Taiwan University Press, 2013), 153–167.

Bekoff, Mark, "The animal's point of view, animal welfare and some other related matters," *Behavioral and Brain Sciences* 14/4 (1991), 753–755.

Bekoff, Marc and Jan Nystrom, "The Other Side of Silence: Rachel Carson's Views of Animals," *Human Ecology Review* 11/2 (2004), 186–200.

Bicknell, Franklin, *The English Complaint or Your Fatigue and its Cure* (London: William Heinemann, 1952).

Bicknell, Franklin, *Chemicals in Food and in Farm Produce: Their Harmful Effects* (London: Faber and Faber, 1960).

Bjørkdahl, Kristian and Tone Druglitrø, eds. *Animal housing and human-animal relations: Politics, practices and infrastructures* (London: Routledge, 2016).

Boyd, William, "Making Meat: Science, Technology, and American Poultry Production," *Technology and Culture* 42/4 (2001), 631–664.

Brace, Catherine, "Looking back: the Cotswalds and English national identity, c. 1890–1950," *Journal of Historical Geography* 25/4 (1999), 502–516.

Bressalier, Michael, Angela Cassidy, Abigail Woods, "One Health in history," in J. Zinsstag et al. (eds.), *One Health: The Theory and Practice of Integrated Health Approaches* (Oxfordshire: CABI, 2015), 1–15.

Brigandt, Ingo, "The instinct concept of the early Konrad Lorenz," *Journal of the History of Biology* 38/3 (2005), 571–608.

Broom, Donald M., "The importance of measures of poor welfare - response to Marian Dawkins "From an Animal's Point of View," *Behavioural and Brain Sciences* 13 (1990), 14–15.

Broom, Donald M., "Welfare as a Broad Scientific Concept," *Journal of Applied Animal Welfare Science* 1/2 (1998), 149–151.

Broom, Donald M., "A History of Animal Welfare Science," *Acta Biotheor* 59 (2011), 121–137.

Broom, Donald M., "Ruth Harrison's Later Writings and Animal Welfare Work," in *Animal Machines - New Edition* (Wallingford and Boston: CABI, 2013), 21–25.

Broom, Donald M., "World Impact of ISAE: past and future," in Jennifer Brown, Yolande Seddon and Michael Appleby (eds), *Animals and Us - 50 years and more of applied ethology* (Wageningen: Wageningen Academic Publishers, 2016), 269–278.

Broom, Donald M. and Andrew F. Fraser, *Domestic animal behaviour and welfare* (Wallingford and Oxford: CABI, 2015).

Brown, Judith M., "'Life Histories' and the History of Modern South Asia," *The American Historical Review* 114/3 (2009), 587–595.

Buller, Henry, "Animal welfare: from production to consumption," in H. Blokhouis et al. (eds.), *Welfare quality: science and society improving animal welfare* (Wageningen: Wageningen Academic Press, 2013), 49–69.

Buller, Henry and Emma Roe, "Modifying and commodifying farm animal welfare: The economisation of layer chickens," *Journal of Rural Studies* 33 (2014), 141–149.

Buller, Henry and Emma Roe, *Food and animal welfare* (London: Bloomsbury Publishing, 2018).

Burdick, Timothy and Pink Dandelion, "Global Quakerism 1920–2015," in Stephen W. Angell and Pink Dandelion (eds.), *The Cambridge Companion to Quakerism* (Cambridge: Cambridge University Press, 2018), 49–66.

Burkett, Jodi, "The Campaign for Nuclear Disarmament and changing attitudes towards the Earth in the nuclear age," *British Journal for the History of Science,* 45/4 (2012), 625–639.

Burkhardt, Richard W., "Founders of Ethology And The Problem Of Human Aggression. A Study In Ethology's Ecologies," in Angela N. H. Creager and William Chester Jordan (eds.), *The Animal/Human Boundary: Historical Perspectives* (Rochester: University of Rochester Press, 2002), 265–304.

Burkhardt, Richard W., *Patterns of Behavior. Konrad Lorenz, Niko Tinbergen, and the Founding of Ethology* (Chicago and London: University of Chicago Press, 2005).

Burkhardt, Richard W., "Tribute to Tinbergen: Putting Niko Tinbergen's 'Four Questions' in Historical Context," *Ethology* 120 (2014), 215–223.

Carbone, Larry, *What animals want: expertise and advocacy in laboratory animal welfare policy* (Oxford: Oxford University Press, 2004).

Carson, Cathryn, "Bildung als Konsumgut: Physik in der westdeutschen Nachkriegskultur," in Dieter Hoffmann (ed.), *Physik im Nachkriegsdeutschland* (Frankfurt: Harri Deutsch, 2003), 73–85.

Carson, Cathryn, "Science as instrumental reason: Heidegger, Habermas, Heisenberg," *Cont Philos Rev* 42 (2010), 483–509.

Carson, Rachel, *Silent Spring* (New York: Houghton Mifflin, 1962).

Cassidy, Angela, *Vermin, victims and disease: British debates over bovine tuberculosis and badgers* (London: Palgrave Macmillan, 2019).

Cederholm, Erika Andersson, Amelie Björck, Kristina Jennbert, and Ann-Sofie Lönngren (eds.), *Exploring the Animal Turn. Human-Animal Relations in Science, Society and Culture* (Lund: Pufendorf Institute for Advanced Studies, 2014).

Chadkirka, James, *Patterns of Membership and Participation among British Quakers, 1823–2012* (Birmingham: MA thesis, University of Birmingham, 2014).

Chase, Malcolm, "This is no claptrap: this is our heritage," in Christopher Shaw and Malcolm Chase (eds), *The Imagined Past: History and Nostalgia* (Manchester: Manchester University Press, 1989), 128–146.

Clark, Brett and John Bellamy Foster, "Henry S. Salt, socialist animal rights activist: An introduction to Salt's A Lover of Animals," *Organization & Environment* 13/4 (2000), 468–473.

Collins, Peter Jeffrey, "The development of ecospirituality among British Quakers," Ecozon@ 2/3 (2011).

Conford, Philip, *The Origins of the Organic Movement* (Edinburgh: Floris, 2001).

Conford, Philip and Patrik Holden, "The Soil Association," in William Lockeretz (ed.), *Organic Farming: An International History* (Wallingford: CABI, 2001), 187–200.

Costall, Alan, "Lloyd Morgan, and the Rise and Fall of Animal Psychology," *Society & Animals* 6/1 (1998), 13–29.

Cox, Graham, Philip Lowe, and Michael Winter, "From State Direction to Self-Regulation: The Historical Development of Corporatism in British Agriculture," *Policy and Politics* 14/4 (1986), 475–490.

Creager, Angela NH and William C. Jordan, eds. *The animal-human boundary: historical perspectives* (Cambridge MA: Harvard University Press, 2002).

Curry, Helen, *Evolution Made to Order: Plant Breeding and Technological Innovation in Twentieth-Century America* (Chicago: University of Chicago Press, 2016).

Dandelion, Pink, *The Quakers: A very Short Introduction* (Oxford: Oxford University Press, 2008).

Darby, William J., "Review, the Poisons in Your Food by William Longgood," *Science* 131/3405 (1960), 979.

Darwin, Charles, *The expression of emotions in animals and man* (London: Murray, 1872).

Dauvergne, Peter and Kate J. Neville, "Mindbombs of right and wrong: cycles of contention in the activist campaign to stop Canada's seal hunt," *Environmental Politics* 20/2 (2011), 192–209.

Davies, Gail F., Beth J. Greenhough, Pru Hobson-West, Robert GW Kirk, Ken Applebee, Laura C. Bellingan, Manuel Berdoy et al. "Developing a collaborative agenda for humanities and social scientific research on laboratory animal science and welfare," *PLoS One* 11/7 (2016), e0158791.

Davies, Gail F., Richard Gorman, Beth Greenhough, Pru Hobson-West, Robert G.W. Kirk, Dmitriy Myelnikov, Alexandra Palmer et al. "Animal research nexus: a new approach to the connections between science, health and animal welfare," *Medical Humanities* 46/4 (2020), 499–511.

Davies, Tegla A., *Friends Ambulance Unit. The Story of the F.A.U. in the Second World War 1939–1945* (London: George Allen and Unwin Limited, 1947).

Davis, Sophia, "Secluded Suffolk: Countryside Writing, c. 1930–1960," *Island Thinking* (2019), 31–71.

Desmond, Adrian and James Moore, *Darwin's Sacred Cause. How a Hatred of Slavery Shaped Darwin's Views on Human Evolution* (Boston and New York: Houghton Mifflin Harcourt, 2009).

Dentinger, Rachel Mason and Abigail Woods, "Introduction to Working Across Species," *History and Philosophy of the Life Sciences* 40/30 (2018), 1–11.

Dickson, Rachel and Sarah MacDougall, "The Whitechapel Boys," *Jewish Quarterly* 51/3 (2004), 29–34.

Duncan, Ian J.H., "The interpretation of preference tests in animal behaviour," *Applied Animal Ethology* 4 (1978), 197–200.

Duncan, Ian J.H., "D.G.M. Wood-Gush Memorial Lecture: An applied ethologist looks at the question "Why?"," *Applied Animal Behaviour Science* 44 (1995), 205–217.

Duncan, Ian J.H., "Thirty Years of Progress in Animal Welfare Science," *Journal of Applied Animal Welfare Science* 1 (2) (1998), 151–154.

Duncan, Ian J.H., "Science-based assessment of animal welfare: farm animals," *Revue scentifique et technique – Office International Epizooties* 24/2 (2005), 483–486.

Duncan, Ian J.H. and Marian Stamp Dawkins, "The problem of assessing 'well-being' and 'suffering' in farm animals," in D. Smidt et al. (eds.), *Indicators relevant to farm animal welfare* (The Hague: Martinus Nijhoff, 1983), 13–24.

Dyhouse, Carol, "Family Patterns Of Social Mobility Through Higher Education In England In The 1930s," *Journal of Social History* 34/4 (2001), 817–842.

Eadie, Edward N., *Understanding Animal Welfare. An Integrated Approach* (Heidelberg et al.: Springer, 2012).

Ekesbo, Ingvar, "Disease incidence in tied and loose housed dairy cattle and causes," *Acta Agriculturae Scandinavica* 15 (Suppl) (1966), 1–74.

Ekesbo, Ingvar, "The Swedish approach," in Council Of Europe (ed.), *Animal Welfare* (Strasbourg: Council of Europe, 2006), 185–197.

Engermann, David C., "Social science in the Cold War," *Isis* 101/2 (2010), 395–399.

Eyton, Audrey, *The Kind Food Guide. Kinder to animals - much kinder to you* (London: Penguin, 1991).

FAWC, *Farm Animal Welfare in Great Britain: Past, Present and Future* (London: FAWC, 2009).

FAWC, *Evidence And The Welfare Of Farmed Animals. Part 1: The Evidence Base* (London: FAWC, 2014).

Finlay, Mark R., "Hogs, Antibiotics, and the Industrial Environments of Postwar Agriculture," in Philip Scranton and Susan R. Schrepfer (eds.), *Industrializing Organisms. Introducing Evolutionary History* (London: Routledge, 2004), 237–260.

Fitzgerald, Deborah, *Every Farm a Factory: The Industrial Ideal in American Agriculture* (New Haven: Yale University Press, 2010).

Food Ethics Council and Heather Pickett, *Farm Animal Welfare, Past, Present and Future* (Southwater: RSPCA, 2014).

Francione, Gary L., *Rain Without Thunder: The Ideology of the Animal Rights Movement* (Philadelphia: Temple University Press, [1996] 2007).

Francione, Gary L. and Anna E. Charlton, "Animal rights," in Linda Kalof (ed.), *The Oxford handbook of animal studies* (Oxford: Oxford University Press, 2017), 25–40.

Fraser, Andrew, "Displacement activities in domestic animals," *British Veterinary Journal* 115 (1959), 195–200.

Fraser, David and Ian J.H. Duncan, "'Pleasures', 'Pains' and Animal Welfare: Toward a Natural History of Affect," *Animal Welfare* 7/4 (1998), 383–396.

Fraser, David, "Animal ethics and animal welfare science: bridging the two cultures - the D.G.M. Wood-Gush Memorial Lecture," *Applied Animal Behaviour Science* 65 (1999), 171–189.

Fraser, David, "Biology of Animal Stress. Implications for Animal Well-being," *Journal of Applied Animal Welfare Science* 2/2 (1999), 157–159.

Fraser, David, "Understanding Animal Welfare," *Acta Veterinaria Scandinavica* 50/Supplement (2008), 1.

Fraser, David, *Understanding Animal Welfare. The Science in its Cultural Context* (Oxford: UFAW, 2008).

Fraser, David, "Ruth Harrison - A Tribute," *Animal Machines - New Edition* (Wallingford and Boston: CABI, 2013), 17–20.

Gaarder, Emily, *Women and the Animal Rights Movement* (New Brunswick: Rutgers University Press, 2011).

Garner, Robert, *Animals, Politics and Morality* (Manchester and New York: Manchester University Press, 1993).

Garner, Robert and Yewande Okuleye, *The Oxford Group and the Emergence of Animal Rights* (Oxford: Oxford University Press, 2020).

Gibbs, Anthony Matthews, *A Bernard Shaw Chronology* (London: Palgrave, 2001).

Gillespie, Neal C., "The Interface of Natural Theology and Science in the Ethology of W. H. Thorpe," *Journal of the History of Biology* 23/1 (1990), 1–38.

Godley, Andrew and Bridget Williams, "Democratizing luxury and the contentious 'invention of the technological chicken' in Britain," *Business History Review* 83/2 (2009), 267–290.

Godley, Andrew, "The emergence of agribusiness in Europe and the development of the Western European broiler chicken industry, 1945 to 1973," *Agricultural History Review* 62/2 (2014), 315–336.

Godlovitch, Roslind, Godlovitch, Stanley, and Harris, John (eds.), *Animals, Men And Morals. An enquiry into the maltreatment of non-humans* (London: Victor Gollancz, 1971).

Grant, Doris, *Housewives Beware* (London: Faber and Faber, 1958).

Grant, Doris, *Your Bread and Your Life* (London: Faber and Faber, 1961).

Gregory, James, *Of Victorians and Vegetarians: The Vegetarian Movement in Victorian Britain* (London and New York: Tauris, 2007).

Griffin, Donald, *The Question of Animal Awareness. Evolutionary Continuity of Mental Experience* (New York: Rockefeller University Press, 1976).

Guenther, Katja, "Monkeys, Mirrors, And Me: Gordon Gallup And The Study of Self-Recognition," *Journal of the History of the Behavioural Sciences* 53/1 (2017), 5–27.

Gradmann, Christoph, *Laboratory Disease: Robert Koch's Medical Bacteriology*, Elborg Forster (trans.), (Baltimore: Johns Hopkins University Press, 2009).

Harraway, Donna J., *When Species Meet* (Minneapolis: University of Minnesota Press, 2007).

Harrison, Ruth, *Animal Machines* (London: Vincent Stuart Ltd, 1964).

Harrison, Ruth, "On Factory Farming," in Roslind Godlovitch, Stanley Godlovitch, and John Harris (eds.), *Animals, Men and Morals. An enquiry into the maltreatment of non-humans* (London: Viktor Gollancz Ltd, 1971), 11–24.

Harrison, Ruth, "Introduction – Proceedings of Workshop sponsored by the Farm Animal Care Trust and the Universities Federation for Animal Welfare: Behavioural needs of Farm Animals," *Applied Animal Behaviour Science*, 19 (1988), 341.

Harrison, Ruth, *Animal Machines - New Edition* (Wallingford and Boston: CABI, 2013).

Helliwell, Richard, Carol Morris, and Sujatha Raman, "Antibiotic stewardship and its implications for agricultural animal-human relationships: Insights from an intensive dairy farm in England," *Journal of Rural Studies* 78 (2020), 447–456.

Herber, Lewis [Pseudonym for Murray Bookchin], *Our Synthetic Environment* (New York: Knopf, 1962).

Heymann, Matthias, "1970s: Turn of an Era in the History of Science?," *Centaurus* 59/1–2 (2017), 1–9.

Hilton, Matthew, James McKay, Nicholas Crowson, and Jean-François Mouhout, *The Politics of Expertise: How NGOs Shaped Modern Britain* (Oxford: Oxford University Press, 2013).

Hinde, R.A., "William Homan Thorpe. 1 April 1902–7 April 1986," *Biographical Memoirs of Fellows of the Royal Society*, 33 (1987), 620–639.

Holderness, B. A., *British agriculture since 1945* (Manchester: Manchester University Press, 1985).

Holland, R.F., *European Decolonization 1918–1981. An Introductory Survey* (1992 edn.; London: Macmillan, 1985).

Holroyd, Michael, *Bernard* Shaw (London: Random House, 2011).

Holroyd, Michael, *Bernard Shaw. 1918–1950. The Lure of Fantasy* (London: Chatto & Windus, 1991).

Horowitz, Roger, "Making the Chicken of Tomorrow. Reworking Poultry as Commodities and as Creatures, 1945–1990," in Susan R. Schrepfer and Scranton Philip (eds.), *Industrializing Organisms. Introducing Evolutionary History* (New York et al., 2004), 215–235.

House of Commons, *First Report from the Agriculture Committee, Animal Welfare in Poultry, Pig and Veal Calf Production, Session 1980–1981, 02.07.1981* (London: House of Commons, 1981).

Howkins, Alun and Linda Merricks, "'Dewy-Eyed Veal Calves'. Live Animal Exports and Middle-Class Opinion, 1980–1995," *The Agricultural History Review* 48/1 (2000), 85–103.

Huxley, Elspeth, *Brave New Victuals. Are We All Being Slowly Poisoned? A Terrifying Enquiry Into The Techniques of Modern Food Production* (London: Panther Books, [1965] 1967).

Huxley, Julian, "Transhumanism," *Journal of Humanistic Psychology* 8/1 (1968), 73–76.

Jasanoff, Sheila, and Kim Sang-Hyun, "Sociotechnical imaginaries and national energy policies," *Science as Culture* 22/2 (2013), 189–196.

Kean, Hilda, *Animal Rights. Political and Social Change in Britain since 1800* (London: Reaktion Books, 1998).

Kean, Hilda, *The Great Cat and Dog Massacre. The Real Story of World War Two's Unknown Tragedy* (Chicago: University of Chicago Press, 2017).

Kean, Hilda and Philip Howell, eds. *The Routledge Companion to Animal-Human History* (London: Routledge, 2018).

Keeling, Linda J., Jeff Rushen, and Ian JH Duncan, "Understanding animal welfare," in Michael C. Appleby, Anna Olsson, and Francisco Galindo (eds), *Animal welfare* (Wallingford and Oxford: CABI, 2018), 13–26.

Kendall, Ena, "Ruth and the ruthless," *The Vegetarian,* /New Series No. 43 (April) (1975), 3 and 21.

Kinkela, David, *DDT and the American Century: Global Health, Environmental Politics, and the Pesticide That Changed the World* (Chapel Hill: University of North Carolina, 2011).

Kirchhelle, Claas, "Swann song: antibiotic regulation in British livestock production (1953–2006)," *Bulletin of the History of Medicine* 92/2 (2018), 317–350.

Kirchhelle, Claas, *Pyrrhic Progress: The History of Antibiotics in Anglo-American Food Production* (New Brunswick: Rutgers University Press, 2020).

Kirk, Robert G.W., *Reliable animals, responsible scientists: constructing standard laboratory animals in Britain c. 1919–1976* (London: PhD Thesis University of London, 2005).

Kirk, Robert G.W., "A brave new animal for a brave new world: The British Laboratory Animals Bureau and the constitution of international standards of laboratory animal production and use, circa 1947–1968," *Isis* 101/1 (2010), 62–94.

Kirk, Robert G.W., "The Invention of the 'Stressed Animal' and the Development of a Science of Animal Welfare, 1947–86," in David Cantor and Edmund Ramsden (eds.), *Stress, Shock, and Adaptation in the Twentieth Century* (Woodbridge and Rochester: University of Rochester Press, 2014), 241–263.

Kirk, Robert G.W., "Recovering the principles of humane experimental technique: The 3Rs and the human essence of animal research," *Science, Technology, & Human Values* 43/4 (2018), 622–648.

Kirk, Robert GW and Edmund Ramsden. "Working across species down on the farm: Howard S. Liddell and the development of comparative psychopathology, c. 1923–1962," *History and philosophy of the life sciences* 40/1 (2018), 1–29.

Kirk, Robert G.W., Neil Pemberton, and Tom Quick. "Being well together? promoting health and well-being through more than human collaboration and companionship," *Medical humanities* 45/1 (2019), 75–81.

Kirk, Robert G.W., "Science and humanity: national culture, scientific freedom and the limits of animal experiment in Britain and America, 1949–1966" – presented at the 2019 LSE/Animal Research Nexus "National Cultures of Care, Animals and Science" workshop, *in preparation*.

Knorr, Karin D. "Tinkering toward Success: Prelude to a Theory of Scientific Practice," *Theory and Society* 8/3 (1979), 347–376.

Kohler, Robert E., *Landscapes and Labscapes: Exploring the lab-field border in biology* (Chicago: University of Chicago Press, 2002).

Kramer, Ann, *Conscientious Objectors of the First World War: a determined resistance* (Barnsley: Pen & Sword, 2014).

Kroll, Garry, "The 'Silent Springs' of Rachel Carson: Mass media and the origins of modern environmentalism," *Public Understanding of Science* 10/4 (2001), 403–420.

Kruuk, Hans, *Niko's Nature. The Life of Niko Tinbergen and his Science of Animal Behaviour* (Oxford: Oxford University Press, 2003).

Lapsansky, Emma Jones, "The Changing World of Quaker Material Culture," in Stephen W. Angell and Pink Dandelion (eds.), *The Cambridge Companion to Quakerism* (Cambridge: Cambridge University Press, 2018), 147–158.

Laurence, Dan H., *Bernard Shaw: Theatrics. Selected Correspondence of Bernard Shaw* (Toronto et al.: University of Toronto Press, 1995).

Lee, Hermione, *Biography: A Very Short Introduction* (Oxford: Oxford University Press, 2009).

Lent, Adam, *British Social Movements Since 1945. Sex, Colour, Peace and Power* (Basingstoke and New York: Palgrave, 2001).
Longgood, William, *The Poisons In Your Food* (New York: Simon and Schuster, 1960).
Lorenz, Konrad, *Das Sogenannte Böse. Zur Naturgeschichte der Aggression* (Wien: Borotha Schoeler, 1963).
Lorimer, Jamie, *Wildlife in the Anthropocene: conservation after nature* (Minneapolis: University of Minnesota Press, 2015).
Lowenthal, David, "British National Identity and the English landscape," *Rural History* 2/2 (1991), 205–230.
Lytle, Mark Hamilton, *The gentle subversive: Rachel Carson, Silent Spring, and the rise of the environmental movement* (Oxford: Oxford University Press, 2007).
MacDougall, Sarah, "Whitechapel Girl: Clare Winsten and Isaac Rosenberg," in Sarah Macdougall, Dickson Rachel, and Ben Uri Art Gallery (eds.), *Whitechapel at war: Isaac Rosenberg & his circle* (London: Ben Uri Gallery 2008), 99–117.
MacDougall, Sarah, "'Something is happening there': Early British modernism, the Great War and the 'Whitechapel Boys'," in Michael J. K. Walsh (ed.), *London, Modernism, and 1914* (Cambridge: Cambridge University Press, 2010), 122–147.
Macfie, H. J. H. and Herbert L. Meiselman, *Food Choice Acceptance and Consumption* (London: Blackie Academic & Professional, 1996).
Martin, John, *The Development of Modern Agriculture. British Farming since 1931* (London et al. : Macmillan & St. Martin's Press, 2000).
McKenzle, Callum C., "The Origins of the British field sports society," *The International Journal of the History of Sport* 13/2 (1996), 177–191.
Mellor, David J., "Moving beyond the 'Five Freedoms' by Updating the 'Five Provisions' and Introducing Aligned 'Animal Welfare Aims," *Animals* 6/10 (2016), 59.
Mench, Joy A., "Thirty Years After Brambell: Whither Animal Welfare Science?," *Journal of Applied Animal Welfare Science* 1/2 (1998), 91–102.
Merricks, Linda, "Green Politics: Animal Rights, Vegetarianism and Naturism," in David Morley and Kevin Robins (eds.), *British Cultural Studies* (Oxford: Oxford University Press, 2001), 431–442.
Millman, Suzanne, Ian Duncan, Markus Stauffacher, and Joseph Stookey, "The impact of applied ethologists and the International Society for Applied Ethology in improving animal welfare," *Applied Animal Behaviour Science* 86 (2004), 299–311.
More-Colyer, R., "Towards 'Mother Earth': Jorian Jenks, Organicism, the Right and the British Union of Fascists," *Journal of Contemporary History* 39/3 (2004), 353–371.
Morris, Desmond, *The Naked Ape. A Zoologist's Study of the Human Animal* (New York: McGraw Hill, 1967).

Myelnikov, Dmitriy, "Cuts and the cutting edge: British science funding and the making of animal biotechnology in 1980s Edinburgh," *The British Journal for the History of Science* 50/4 (2017), 701–728.

Myelnikov, Dmitriy, "Tinkering with genes and embryos: the multiple invention of transgenic mice c. 1980," *History and Technology* 35/4 (2019), 425–452.

Nasaw, David. "AHR Roundtable: Historians and Biography," *American Historical Review* 114 (2009), 573.

Nelson, Nicole C., *Model behavior: Animal experiments, complexity, and the genetics of psychiatric disorders* (Chicago: University of Chicago Press, 2018).

Newberry, R. C. and Victoria Sandilands, "Pioneers of applied ethology," in *Animals and us: 50 years and more of applied ethology* (Wageningen: Wageningen Academic Publishers, 2016), 175–192.

Nixon, Sean, "Trouble at the National Trust: Post-war Recreation, the Benson Report and the Rebuilding of a Conservation Organization in the 1960s," *Twentieth Century British History* 26/4 (2015), 529–550.

Overy, Richard, "Pacifism and the Blitz, 1940–1941," *Past & Present* 219 (2013), 201–219.

Palmer, Clare and Peter Sandøe, "Animal ethics," in Michael C. Appleby, I. Anna S. Olsson, Francisco Galinda (eds.), *Animal Welfare* (Wallingford: CABI, 2018), 3–15.

Parkin, Frank, *Middle Class Radicalism: The Social Bases of the British Campaign for Nuclear Disarmament* (Manchester: Manchester University Press, 1968).

Patterson, Ian, *The Translation of Soviet Literature* (Oxford: Oxford University Press, 2013).

Pemberton, Neil and Mike Worboys, *Rabies in Britain. Dogs, Disease and Culture, 1830–2000* (London: Palgrave Macmillan, 2006).

Petherick, J.C. and Ian J.H. Duncan, "The International Society for Applied Ethology: going strong 50 years on," in Jennifer Brown, Yolande Seddon and Michael Appleby (eds), *Animals and Us - 50 years and more of applied ethology* (Wageningen: Wageningen Academic Publishers, 2016), 27–50.

Preece, Rod, *Awe for the Tiger, Love for the Lamb: A Chronicle of Sensibility to Animals* (London and New York: Routledge, 2002).

Preece, Rod, *Animal Sensibility and Inclusive Justice in the Age of Bernard Shaw* (Vancouver: UBC Press, 2011).

Radick, Gregory, "Animal agency in the age of the Modern Synthesis: W.H. Thorpe's example," *British Journal of the History of Science Themes* 2 (2017), 35–56.

Radkau, Joachim, *Die Ära der Ökologie. Eine Weltgeschichte* (München: C.H. Beck, 2011).

Rasmussen, Nicolas, "Goofball Panic: Barbiturates, 'Dangerous' and Addictive Drugs, and the Regulation of Medicine in Postwar America," in Jeremy A. Greene and Elizabeth Siegel Watkins (eds.), *Writing, Filing, Using, and Abusing the Prescription in Modern America* (Baltimore: Johns Hopkins University Press, 2012), 23–45.

Report of the Technical Committee to Enquire into the Welfare of Animals kept under Intensive Livestock Systems (London: HMSO, 1965).

Richardson, Elsa, "Man is not a meat-eating animal: vegetarians and evolution in late-Victorian Britain," *Victorian Review* 45/1 (2019), 117–134.

Ritvo, Harriet, *The Animal Estate. The English and Other Creatures in Victorian England* (Cambridge MA: Harvard University Press, 1987).

Rollin, Bernard E., "Animal Production and the new social ethic for animals," *Journal of Social Philosophy* 25th special issue (1994), 71–83.

Rollin, Bernard E., "Animal Machines - Prophecy and Philosophy," *Animal Machines - New Edition* (Wallingford and Boston: CABI, 2013), 10–16.

Romanes, John George, *Animal Intelligence* (New York: D. Appleton, [1878] 1884).

Romanes, John George, *Mental evolution in animals* (Kegan Paul, Trench, 1883).

Roscher, Mieke, *Ein Königreich für Tiere. Die Geschichte der britischen Tierrrechtsbewegung* (Marburg: Tectum Verlag, 2009).

Russell, Edmund, *War and Nature: Fighting Humans and Insects with Chemicals from World War I to "SilentSpring"* (Cambridge: Cambridge University Press, 2001).

Ryder, Richard D., "Experiments on Animals," in Roslind Godlovitch, Stanley Godlovitch, and John Harris (eds.), *Animals, Men and Morals* (London 1971, 1971), 41–82.

Ryder, Richard D., "Putting Animals into Politics," in Robert Garner (ed), *Animal Rights. The Changing Debate* (Basingstoke and London: Macmillan, 1996), 166–193.

Ryder, Richard D., *Animal Revolution. Changing Attitudes Towards Speciesism* (Oxford and New York: Berg, 2000).

Ryder, Richard D., "Harrison, Ruth (1920–2000)," *Oxford Dictionary of National Biography* (Oxford University Press, 2004).

Ryder, Richard D., "RSPCA Reform Group," in Marc Bekoff and Carron A. Meaney (eds.), *Encyclopedia of Animal Rights and Animal Welfare* 2nd Edition (Abingdon/ New York: Routledge, {1998} 2013), 307–308.

Sainsbury, David, *Farm Animal Welfare. Cattle, Pigs and Poultry* (London: Collins, 1986).

Sams, Craig, "Introduction," in Simon Wright (ed), *Handbook of Organic Food Processing and Production* (Dordrecht: Springer Science + Business Media 1994), 1–16.

Sandilands, Victoria, "David Wood-Gush The Biography of an Ethology Mentor," *Applied Animal Behaviour Science* 87 (2004), 173–176.

Sayer, Karen, "Animal Machines: The Public Response to Intensification in Great Britain, c. 1960–c. 1973," *Agricultural History* 87/4 (2013), 473–501.

Sayer, Karen, *Farm Animals in Britain*, 1850–2001 (New York: Taylor & Francis, 2018).

Sen, Sudipta, *Ganges: The Many Pasts of an Indian River* (New Haven: Yale University Press, 2019).

Shapin, Steven and Simon Schaffer, *Leviathan and the air-pump: Hobbes, Boyle, and the experimental life* (Princeton: Princeton University Press, [1985] 2011).

Sheldon, Sally, Gayle Davis, Jane O'Neill, and Clare Parker. "The Abortion Act (1967): a biography," *Legal Studies* 39/1 (2019), 18–35.

Silverman, Chloe, "'Birdwatching and baby-watching': Niko and Elisabeth Tinbergen's ethological approach to autism," *History of psychiatry* 21/2 (2010), 176–189.

Simon, Christian, *DDT. Kulturgeschichte einer Chemischen Verbindung* (Basel: Christian Merian Verlag, 1999).

Singer, Peter, *Animal Liberation. A New Ethics for Our Treatment of Animals.* Harper Collins (New York) 1975.

Singer, Peter, "Animal Liberation: A Personal View," *Between the Species* (1986), 148–154.

Singer, Peter, *Animal Liberation: A Personal View. Writings on an Ethical Life* (London: Fourth Estate, 2001).

Smith, Michael B., "'Silence, Miss Carson!' Science, Gender, and the Reception of 'Silent Spring'," *Feminist Studies* 27/3 (2001), 733–752.

Smith, Roger, "Biology and values in interwar Britain: CS Sherrington, Julian Huxley and the vision of progress," *Past & Present* 178 (2003), 210–242.

Sparks, John, *The Discovery of Animal Behavior* (London: William Collins & Co, 1982).

[Stamp] Dawkins, Marian, "Do hens suffer in battery cages? Environmental preferences and welfare," *Animal Behaviour* 25 (1977), 1034–1046.

[Stamp] Dawkins, Marian, *Animal Suffering. The Science of Animal Welfare* (London and New York: Chapman and Hall, 1980).

Stamp Dawkins, Marian, "From an animal's point of view: Motivation, fitness, and animal welfare," *Behavioural and Brain Sciences 13* (1990), 1–9.

Stamp Dawkins, Marian, "Why has there not been more progress in animal welfare research? - D.G.M. Wood-Gush Memorial Lecture," *Applied Animal Behaviour Sciences* 53 (1997), 59–73.

Stamp Dawkins, Marian, "Why We Still Need to Read *Animal Machines*," in *Animal Machines - New Edition* (Wallingford and Boston: CABI, 2013), 1–4.

Stolba, Alex & D.G.M. Wood-Gush, "The identification of behavioural key features and their incorporation into a housing design for pigs," *Annales des Recherches Véterinaire* 15 (1984), 287–298.

Swart, Sandra, "The other citizens: Nationalism and animals," in Hilda Kean and Philip Howell (eds), *The Routledge Companion to Animal-Human History* (London: Routledge, 2018), 31–52.

Tessari, Alessandra and Andrew Godley, "Made in Italy. Made in Britain. Quality, brands and innovation in the European poultry market, 1950–80," *Business History* 56/7 (2014), 1057–1083.

Tester, Keith, "The British Experience of the Militant Opposition to the Agricultural Use of Animals," *Journal of Agricultural Ethics* 2 (1989), 241–251.

Thomas, Keith, *Man and the natural world: Changing attitudes in England 1500–1800* (London: Penguin UK, 1991).

Thorpe, William Homan, "Zoology and Behavioural Sciences," *Nature* 216 (07.10.1967), 20.

Thorpe, William Homan, "Welfare of Domestic Animals," *Nature* 224 (04.10.1969), 18–20.

Trees, Alexander, "Non-stun slaughter: the elephant in the room," *Veterinary Record* 182/7 (2018), 177.

Tsing, Anna Lowenhaupt, *The mushroom at the end of the world: On the possibility of life in capitalist ruins* (Princeton: Princeton University Press, 2015).

Twigg, Julian, *The Vegetarian Movement in England, 1847–1981: A Study In The Structure Of Its Ideology* (London: Dissertation London School of Economics) 1981.

Uekötter, Frank and Amir Zelinger, "Die Feinen Unterschiede. Die Tierschutzbewegung und die Gegenwart der Geschichte," in Herwig Grimm and Carola Otterstedt (eds.), *Das Tier an sich. Disziplinenübergreifende Perspektiven für neue Wege im wissenschaftsbasierten Tierschutz* (Göttingen: Vandenhoeck & Ruprecht, 2012), 119–134.

UFAW, *The UFAW Handbook on the Care and Management of Farm Animals* (Edinburgh: Churchill Livingston, 1971).

van de Weerd, Heleen and Sandilands, Victoria, "Bringing the issue of animal welfare to the public: A biography of Ruth Harrison (1920–2000)," *Applied Animal Behaviour Science* 113 (2008), 404–410.

Veldman, Meredith, *Fantasy, the Bomb and the Greening of Britain. Romantic Protest, 1945–1980* (Cambridge: Cambridge University Press, 1994).

Wearing, J.P., *Bernard Shaw and Nancy Astor. Selected Correspondence of Bernard Shaw* (Toronto et al.: University of Toronto Press, 2005).

Webster, John, *Animal Welfare. A Cool Eye Towards Eden. A constructive approach to the problem of man's dominion over the animals* (Oxford et al.: Blackwell Science, [1995] 2007).

Webster, John, "Ruth Harrison - Tribute To An Inspirational Friend," in *Animal Machines - New Edition* (Wallingford and Boston: CABI, 2013), 5–9.

Wilson, Dolly Smith, "A New Look at the Affluent Worker: The Good Working Mother in Post-War Britain," *Twentieth Century British History* 17/2 (2006), 206–229.

Wilson, Jean Moorcroft, *Isaac Rosenberg: The Making of a Great War Poet: A New Life* (Chicago: Northwestern University Press, 2009).

Winsten, Stephen, *Days with Bernard Shaw* (London: Readers Union/ Hutchinson, 1951).

Winsten, Stephen, *Salt and His Circle* (London: Hutchinson & Co. Ltd, 1951b).

Winter, Michael, "Corporatism and agriculture in the UK: the case of the milk marketing board," *Sociologia Ruralis* 24/2 (1984), 106–119.

Winter, Michael, *Rural Politics: Policies for Agriculture, Forestry and the Environment* (London: Routledge, 1996).

Wood-Gush, D.G.M., "Animal Welfare in Modern Agriculture," *British Veterinary Journal* 129 (1973), 173.

Woods, Abigail, "From cruelty to welfare: the emergence of farm animal welfare in Britain, 1964–71," *Endeavour* 36/1 (2012), 14–22.

Woods, Abigail, "Is Prevention Better than Cure? The Rise and Fall of Veterinary Preventive Medicine, c. 1950–1980," *Social History of Medicine* 26/1 (2012), 113–131.

Woods, Abigail, "Rethinking the History of Modern Agriculture: British Pig Production, c. 1910–65," *Twentieth Century British History* 23/2 (2012), 165–191.

Zayed, Yago, "Agriculture: historical statistics," *House of Commons Library Briefing Paper* 03339 (2016).

Zelko, Frank, *Make It A Green Peace! The rise of countercultural environmentalism* (Oxford: Oxford University Press, 2013).

Zuckerman, Solly, "The Human beast," *Nature* 212 (05.11.1966), 563-564.

Zweiniger-Bargielowska, Ina, *Austerity in Britain: Rationing, Control and Consumption 1939–1955* (Oxford and New York: Oxford University Press, 2002).

Index[1]

A
Agricultural Research Council (ARC), 134, 185, 185n53
Animal Liberation Front (ALF), 173
Animal Machines, x, 1, 6–9, 11, 13, 14, 16, 17, 46, 48, 78–91, 95–124, 131, 136, 150, 152, 154, 169, 171, 172, 175, 199, 202, 205, 207, 219–221, 240, 242, 243, 245
Animal rights, ix–xi, 7, 9, 16, 150, 167, 170, 173, 201, 211, 241, 243, 244
Animal welfare science, 7, 14, 174–202, 223–225, 229, 232, 239, 241, 244
Anthropomorphism, 11, 53, 56–58, 109, 235
Assurance Schemes, 2, 4, 9, 16, 223, 226, 228, 241, 246, 247

B
Baker, John R., 84, 129
Barter, Gwendolen (Gwen), 80, 81, 156
Behaviour
  affective states, 63, 179, 181, 233, 242
  cognition, 2, 60, 63, 179
  hydraulic model, 55
  innate behaviour, 54, 55
  instincts, 54, 55, 60, 110
  stress, 16, 182, 188, 190, 223, 244
Behaviourism, 53
Birnberg, Clara, *see* Winsten, Clare
Birnberg, Johanes (Jonas), 45
Blood sports, 6, 15, 32, 67, 76, 81, 149, 156, 160, 164, 166, 168
Brambell Committee (Technical Committee to Enquire Into The Welfare Of Animals Kept Under Intensive Livestock Husbandry Systems), x, 14, 107, 108, 110, 111, 114, 117, 120, 122, 128, 129, 138, 143, 155, 171, 175, 243

[1] Note: Page numbers followed by 'n' refer to notes.

C. Kirchhelle, *Bearing Witness*, Palgrave Studies in the History of Social Movements,
https://doi.org/10.1007/978-3-030-62792-8

Brambell Report, 110, 116, 118, 120, 123, 128, 134, 142, 143, 176, 181
Brambell, Rogers, 8, 107, 116, 118–122, 128–130, 137, 143, 153, 179, 208
British Field Sports Society (BFSS), 76, 156–158, 160, 167, 202, 243
British Veterinary Association (BVA), 87, 114, 188, 194
Broom, Donald Maurice, 173, 181, 182, 214, 225, 231, 232, 235, 239
Bryant, John, 162, 169n118
Burden, Frederick Frank Arthur, 139, 155, 157, 158, 160, 164, 166, 168, 188

C
Calves, *see* Veal
Campaign for Nuclear Disarmament (CND), 6, 11, 45, 48, 242
Carson, Rachel, 1, 4, 5, 8, 13, 70, 79, 85–89, 91, 98, 99, 219, 240, 242, 246
Chance, Michael, 58
Corporatism, 146, 226
Council of Europe, 143, 144, 212
Countryside, 28, 68, 69, 74, 75, 78, 79, 81, 97, 98, 114
Crusade Against All Cruelty to Animals, 46, 80, 97

D
Dawkins, Marian Stamp, 182, 183, 183n47, 191–194, 214, 224, 230, 231, 235, 239
Dowding, Muriel, 150
Dugdale, John, 72, 73, 77, 102
Duncan, Ian, 183, 190, 233, 235

E
Ekesbo, Ingvar, 140, 176
Environmentalism, 6, 45, 113, 113n118, 160, 163, 242
Ethology, 7, 11, 12, 51–63, 85n22, 91, 113n118, 133, 175, 177–181, 187, 192, 224, 244
European Economic Community (EEC), 117n147, 128n4, 141, 207, 212, 245
Evolution, 12, 31, 37, 52, 56, 59–62, 178, 219, 241, 242
Ewbank, Roger, 188, 190

F
Factory farming, *see* Intensive farming
Farm Animal Care Trust (FACT), 141n74, 151, 151n11, 152, 152n19, 152n20, 174–202, 205, 213–215, 213n44, 217, 218, 218n71, 220, 239, 240, 244, 246
Farm Animal Welfare Advisory Committee (FAWAC), 15, 118–121, 123, 123n188, 124, 127–146, 151, 153–155, 153n24, 154n26, 167, 172, 174, 176, 177, 179, 187, 195, 197, 199–202, 205, 208, 209, 211, 214, 240, 243–245
Farm Animal Welfare Council (FAWC), 16, 201, 205–215, 209n17, 217–219, 236, 245, 246
Field sports, *see* Blood sports
Five freedoms, 206, 212, 215, 235, 245
Fraser, Andrew, 180, 188, 190, 239
Fraser, David, 183, 196, 224, 233–236
Friends Ambulance Unit (FAU), 10, 38–42, 38n15, 48, 241

G

Greenpeace, 11, 45, 46, 81, 149
Griffin, Donald, 182

H

Harrison, Dexter (Dex), 10, 43, 44, 47n51, 82, 105, 119, 120, 152n19, 200, 217
Harrison, Ruth, x, xi, 1–11, 13–17, 21, 35–48, 38n15, 60, 63, 78–81, 83–91, 85n22, 89n51, 95–124, 127–146, 150–160, 152n20, 162, 167–174, 176, 185, 191, 195–200, 195n122, 197n131, 197n132, 202, 205, 205n1, 207, 210, 213–221, 227, 228, 236, 239–247
Hewer, Humphrey Robert, 127, 128, 130, 132, 139, 144, 145
Hobhouse, Richard John, 156, 158, 160, 161, 163–166, 168, 190, 191
Hughes, Cledwyn, 128, 131, 139
Huxley, Elspeth, 113, 113n118, 120
Huxley, Julian, 10, 12, 54, 56, 61–63, 113n118, 129, 130, 159, 235, 242

I

Intensive farming, 9, 46, 68, 70, 75–77, 81, 83, 85, 88–90, 97, 98, 100, 101, 104, 108, 111–115, 146, 149–151, 153, 155, 167, 170, 186, 188, 193, 208, 225, 240, 246
International Society for Applied Ethology (ISAE), 180, 184, 190, 196, 224

J

Jewell, Peter, 191, 193
Jewish, 10, 11, 22, 24, 26, 42, 87, 150, 154

L

League Against Cruel Sports (LACS), 76, 156, 158
Lorenz, Konrad, 54–57, 59, 60, 84, 175, 177, 178

M

Mench, Joy, 232, 233
Ministry of Agriculture Fisheries and Food (MAFF), 72, 77, 81, 101–108, 111, 116–124, 127–129, 131–133, 136, 140, 141, 143–146, 154n26, 156, 185, 185n53, 187, 194, 198, 201, 206–209, 211, 226, 243–246
Morris, Desmond, 129, 178

N

Napier, John, 186–192
National Farmers' Union of England and Wales (NFU), 66, 75, 98, 99, 114, 150, 194, 228
Nation of Animal Lovers, 13, 65–78
Nerina, Nadia, 159, 169
1911 Protection of Animals Act, 65, 80, 84, 161, 188
1968 Agriculture (Miscellaneous Provisions) Bill, 123, 127, 154
1976 European Convention for the Protection of Animals Kept for Farming Purposes, 142, 142n83

O
Organic, 69, 70, 223, 226, 246
Oxford Group, 9, 170

P
Pacifism, ix, 21, 27, 37
Parliamentary Animal Welfare Group, 71–72, 154, 167
Pigs, 27, 67, 68, 73, 82, 90, 96, 99, 110, 111, 113, 117, 123, 127, 128, 132, 138, 168, 181, 187, 189, 206, 211, 212, 214, 215, 217, 218, 225, 228, 241
Poultry, 68, 81, 90, 102, 104, 111, 128, 132, 134, 139, 140, 153, 154, 191, 197, 215, 217, 225, 228

Q
Quaker, 2, 10, 11, 35, 37, 38, 41, 45–47, 242

R
Regan, Tom, 7, 15, 173, 234
Rollin, Bernard, 234
Rothschild, Miriam, 214, 215
Royal Society for the Prevention of Cruelty to Animals (RSPCA)
  RSPCA Farm Livestock Advisory Committee (FLAC), 163, 168, 186–194, 196, 201, 244
  RSPCA Freedom Foods, 226–229
  RSPCA Reform Group, 16, 149, 150, 160–168, 173, 201, 202, 208
Russell, William Moy Stratton, 59
Ruth Harrison Research Trust, *see* Farm Animal Care Trust (FACT)
Ryder, Richard, 4, 7, 8, 15, 162, 166, 168–172, 191–194, 209, 210, 212, 244

S
Sainsbury, David, 84, 99, 134, 151, 152, 177, 195, 200, 225, 239
Salt, Henry, 31, 32, 66
Seager, Bryan, 162, 164, 170
Shaw, George Bernard, 10, 11, 29, 31–34, 36, 42
Singer, Peter, 7, 15, 170–172, 234
Slaughter, 25, 58, 65, 72, 75, 82, 87, 88, 90, 100, 103, 141, 150, 187, 194, 208, 212, 214, 217, 220, 235
Soames, Christopher, 103, 105–107
Society for Veterinary Ethology, *see* International Society for Applied Ethology
Society of Friends, *see* Quaker
Sows, 131, 132, 136, 138, 181, 187, 189, 211, 213, 215, 228, 229, 240, 241
Standing Committee of the European Convention for the Protection of Animals Kept For Farming Purposes (T-AP), 213, 220, 246
State Veterinary Service (SVS), 132, 139, 141, 187
Stevens, Christine Gesell, 87
Stolba, Alex, 213–214
Stress, *see* Behaviour

T
Thatcher, Margaret, 207–209, 245
Thorpe, William Homan, 10, 12, 14, 37, 38, 59–62, 85, 85n22, 110, 116, 122, 129, 153, 177–179, 181, 182, 191, 218, 235, 242

Thrift, 9, 14, 71, 72, 84, 91, 109, 114, 115, 134, 181
Tinbergen, Nikolaas (Niko), 55–57, 60, 63, 129, 130, 140, 175, 177, 178, 182, 231

U

Universities Federation for Animal Welfare (UFAW), 12, 57–59, 83, 84, 115, 195, 213, 224, 235

V

Veal, xi, 17, 71–75, 77, 80–82, 85, 86, 90, 96, 97, 99, 100, 111, 113, 117, 122, 127, 128, 130, 133–135, 139, 154, 171, 187, 189, 197–200, 211, 212, 215, 219, 220, 240, 241, 245
Vegetarianism, ix, 21, 25, 28, 31, 32, 37, 97, 171, 172

W

Walker, Peter, 209, 209n20
Webster, Anthony (John), 196–200, 199n143, 206, 211, 224, 229, 230, 235, 240
Weinstein, Rachmiel (Aaron), 26
Weinstein, Samuel or Simy, *see* Winsten, Stephen
Whitechapel Boys, 10, 21, 22, 23n9, 24, 26, 28–29, 34
Winsten, Christopher Blake, 31
Winsten, Clare, 10, 21, 22, 22n5, 23n9, 25, 27, 29, 33n52, 34
Winsten, Ruth, *see* Harrison, Ruth
Winsten, Stephen, 10, 21, 26–28, 32–34, 32n50, 36, 104
Winsten, Theodora, 28, 30, 33, 34, 42, 43
Wood-Gush, David G. M., 180–183, 189, 196, 213, 230

Z

Zuckerman, Solly, 107, 133n32, 178